Studies in Chemistry No. 7

Editors: Bryan J. Stokes and
Anthony J. Malpas

The Rusting of Iron : Causes and Control

ULICK R. EVANS,
F.R.S., Sc.D.,

*Emeritus Reader in the Science of Metallic Corrosion
Cambridge University; Honorary Fellow, King's College,
Cambridge*

Appendix adapted from Chapters 14 and 15,
Experimental Metallurgy by D. Eurof Davies, 1966,
Elsevier Publishing Company.

Edward Arnold

© Ulick R. Evans, 1972
First published, 1972
by Edward Arnold (Publishers) Limited,
25 Hill Street,
London, W1X 8LL

Boards edition ISBN: 0 7131 2330 3
Paper edition   ISBN: 0 7131 2331 1

Printed in Great Britain
by Richard Clay (The Chaucer Press) Ltd.,
Bungay, Suffolk

Editors' Preface

Chemistry courses for schools, technical colleges and universities are sometimes criticized on the grounds that they disregard the applied aspects of the subject, and new syllabuses tend to place increasing emphasis on academic principles to the exclusion of industrial practice.

It is, of course, wholly desirable that special attention should be given to the basic principles of the subject, but the influence of chemistry upon industry, medicine, agriculture and indeed the whole structure of society is both immense and increasing. No chemical education should be thought complete without some informed acquaintance with the impact of chemistry on society, and, if the benefits of scientific progress are to continue, capable students must be stimulated to take up careers in applied fields.

One of the aims of the Studies in Chemistry series of booklets is to stimulate students' interest in the application of the chemical principles learned in the school sixth form or technical college to practical problems of modern life. This book deserves special attention in this respect, for here school and college students have the rare opportunity of reading a book on corrosion written specially for them by one of the world's greatest authorities on the subject. Dr. Ulick Evans brings a lifetime of experience to the writing of this book, which is an authoritative, up-to-date account of the present state of our knowledge about the rusting of iron. As such, it takes the student from sixth form chemistry to the frontiers of knowledge in this particular field.

1972   B. J. S.
                                                                      A. J. M.

Preface

Five years ago, I received a kind invitation from the Chemical Society to contribute an article on the 'Mechanism of Rusting'. I agreed, and in due course, my article was published in *Quarterly Reviews* 1967, *21*, No. 1, p. 29. Two authorities on School Science, reading it, independently reached the opinion that the article, slightly modified, would appeal to members of Sixth Forms. The Society kindly agreed to reprinting in booklet form.

The idea that the Mechanism of Rusting, as a subject, would appeal to Sixth Formers was probably a sound one; some reasons for my agreement with the two authorities are suggested later in this Preface. When, however revision was started, it became evident that something far exceeding slight modification was necessary. In the final version, only about one third represents the original *Quarterly Reviews* article.

The point is worthy of emphasis, since the idea that something intended for one purpose can be adapted to another is today a not infrequent source of inefficiency. There are many instruments on the market, skilfully designed for a certain purpose, carefully constructed and, it must be added, extremely expensive. Sometimes one of these is purchased for use in a research which the original designer had not in mind. Doubtless, after drastic modification and costly additions, the purpose is achieved; often, however, it would have saved time and money—besides providing more satisfactory results—had an *ad hoc* apparatus been designed for the research in question.

What is true of apparatus applies equally to books. It implies no criticism either of the young or the old to recognize that their requirements are different. The first class probably displays greater readiness to accept ideas and has less to unlearn than the second; the second class, however, possesses a greater amount of background knowledge and a wider experience— a real advantage if not accompanied by inflexibility of thinking.

In any case, the members of the Sixth Form of today, who may be the University Students of next year, and the holders of key positions in the Country some years later, deserve something that is tailor-made to their requirements; they should not be treated on the lines of the old song, 'Father's breeks will soon fit Willy'.

The rusting of iron has special claim to the interest of school pupils— since, unlike most of the chemical reactions studied at school, they see it in their ordinary lives; it is not something produced with strange laboratory apparatus during the periods assigned to 'practical chemistry'; indeed it is a change familiar to those who receive no laboratory training at all. It is, however, a more difficult subject than most of those chosen for study in school hours; even proficient scientists have found difficulty in understanding it, and few subjects have occasioned more violent controversy. Some readers of this book may find it a Challenge, but if they accept the Challenge in the spirit exhibited on the sports ground, they should (perhaps, with help from their masters) overcome the difficulties and derive satisfaction from the victory. After all, someone who is not prepared to face difficulties is unlikely to become a great scientist.

I have endeavoured to keep the main text as simple as possible; in a few places, I have introduced some less easy features which may arouse interest although they are not essential to the main argument. These have been placed in footnotes, which the less advanced reader may prefer to omit or to defer to a second reading of the book; an example is the long footnote on

p. 3, referring to Oxidation. Another use of footnotes has been to explain terms with which a few readers may be unfamiliar; such explanation, if placed in the text, would interrupt the argument and possibly annoy the more advanced reader; an example is the footnote on p. 6. dealing with anodic and cathodic processes. It is hoped that this policy will have provided a discussion which can be understood by most readers. That the treatment is suitable for schools is the opinion expressed by some who have more experience of school teaching than the Author. I would thank A. J. Malpas, one of the General Editors of the series, who has made some valuable suggestions, and also my friend, D. H. Peacock, who has kindly looked through the draft version of the book and whose encouragement has been greatly appreciated.

It was suggested by the General Editors that the inclusion of some experiments would increase the usefulness of the book. I felt doubtful about introducing experiments which had not been thoroughly tested out with the materials available today. Then I remembered that my friend and former colleague, Dr. D. Eurof Davies, had worked out a course of experiments in metallurgy which had been proved to work in practical classes at Swansea. These are described in his book, 'Practical Experimental Metallurgy', published by Elsevier. Two chapters in the book contain experiments on Corrosion. By the kind approval of Dr. Eurof Davies, and the generous consent of the Elsevier Publishing Company (owners of the Copyright), these two chapters are reproduced in the present volume—with only slight alteration agreed between Dr. Eurof Davies and myself. I feel that this co-operation has added greatly to the value of the book.

Whilst the book was being written, an experimental study of the mechanism of atmospheric rusting was being carried out, under my supervision, by C. A. J. Taylor, working in the Department of Metallurgy, Cambridge University, through the kind permission of Professor R. W. K. Honeycombe and Dr. J. E. O. Mayne. The research is supported by the Department of Trade and Industry (formerly Ministry of Technology). Some of Mr. Taylor's curves – which have already been published in a journal – are reproduced in this book; his latest results, not yet ready for publication, have been of great value in showing that the opinions expressed in the book are essentially correct.* In expressing my indebtedness to Mr. Taylor's skill and perseverance, I would add my thanks to the Department of Trade and Industry and especially to Mr. H. G. Cole, whose help and encouragement have been greatly appreciated.

1972   U. R. E.

* Since these lines were written, publication has taken place; see U. R. Evans and C. A. J. Taylor, *Corrosion Science* 1972, **12**, 227.

Contents

1 Introduction

1.1 Practical importance of Corrosion

Some years ago it was customary to start a lecture on Corrosion by stating that the annual cost of Corrosion Damage and Preventive Measures in the U.K. was about £600 000 000; this estimate—due to the late Dr. W. H. V. Vernon, a recognized Authority—duly shocked the audiences of that time. Today people are less easily shocked; they have become accustomed to the quotation of astronomical figures for the cost of Space Projects or Defence Measures. If, however, it is remembered that the figure quoted exceeds the Adverse Trade Balance of the Country in a bad year—a balance which has been the cause of wage restrictions and other unpopular measures—it becomes clear that measures to reduce corrosion damage might really be beneficial to national welfare.

Some examples may be given of the forms which trouble caused by Corrosion may take. Certain natural waters, if conveyed through copper pipes, can acquire traces of copper salts by a very slow (and, in itself, harmless) corrosion process. The damage to the copper pipes is trivial, but if the water then passes into galvanized iron tanks, traces of metallic copper are deposited, and the corrosion-couple zinc-copper is set up, which may soon destroy the protective zinc covering; after that, the couple iron-copper leads to attack upon the iron and leakage may develop. This can occur whether or not the copper and galvanized iron are in electrical connection.

Kenworthy[32] has studied trouble set up in that way on housing estates. On one estate, where copper pipes were installed with galvanized tanks, *half* the installations had failed in four years, although on another estate *using the same water* there were *no failures*, because here both pipes and tanks were of galvanized iron. On two other estates, situated close together, but using different waters, copper pipes and galvanized tanks were fitted; on one estate *every* installation had failed after four years, since the water was one which dissolved a considerable amount of copper; on the other estate, there were *no failures* to the tanks even after ten years, because the water was one which dissolved practically no copper.

Our national papers often report a trouble described in Fleet Street as *Metal Fatigue*. Sometimes it is stated that a fleet of aircraft have been 'grounded' as a precaution, owing to the discovery of cracks on one of them;

or perhaps it may be announced that machinery has failed or a structure collapsed. More often than not the trouble should more accurately have been described as *Corrosion Fatigue*. It is true that fatigue cracking can occur in absence of Corrosion. Everyone knows that, by sharply bending a strip of metal sheet, or a piece of wire, first in one direction and then in another, it is often possible to make the metal break. That is an example of pure fatigue, in which Corrosion has played no part. But pure fatigue only causes failure if the stress applied exceeds a certain limit, the value of which is known for most structural materials; thus, in absence of Corrosion, it is possible to make mathematical calculations and to ensure that our aircraft or machines are operating under absolutely safe conditions. But in a corrosive environment failure can occur under conditions which, in absence of Corrosion, would enable working to continue safely for an indefinite period. Calculation then becomes very difficult.

Another closely related type of failure is known as *Stress-Corrosion Cracking*. This occurs under steady tensile stress, whereas Corrosion Fatigue arises from stress which is alternating or fluctuating. In contrast to Corrosion Fatigue which can affect nearly all materials, Stress-Corrosion Cracking only affects a limited number of materials—and in some cases only affects them if they have received unsuitable heat-treatment; a susceptible material, stressed in a corrosive environment, may fracture even though the stress applied is far below the value of the Ultimate Tensile Stress (so that, in absence of Corrosion, there would have been no risk of failure). Materials which can fail by Stress-Corrosion Cracking include several aircraft alloys and austenitic stainless steel—an alloy well known in everyday life.

1.2   Scope of this book

Because the rusting of iron is a corrosion process likely to be familiar to the reader, and also one which is well understood, this little book will confine itself to that type of Corrosion—although on pages 31–34 a few words will be found about the behaviour of some other metals in a corrosive environment, especially those used in providing protective coverings for iron.

There is no reason to avoid the use of iron or steel as structural material merely through fear of rusting. If suitable paint-coats are applied at appropriate intervals of time, there need generally be no weakening of a steel structure due to Corrosion. But the paint-system must be chosen correctly, and must be designed to bring substances which inhibit corrosion reactions into contact with the metallic surface. It is commonly believed that paints protect by providing a barrier which excludes corrosive substances from the metal. That is, in general, untrue; many excellent paints are pervious to corrosive substances; they act by suppressing the dangerous reactions. The principles governing the prevention of rusting by paint are briefly discussed on pp. 34–37.

1.3 Scaling and rusting

The rusting of iron (the formation of hydrated oxide in presence of oxygen and water) must be distinguished from scaling (the formation of anhydrous oxide). In *scaling*, the oxidation-rate falls off as the thickness of the oxide-scale increases. When iron is exposed to air at a high temperature the thickening-rate of the scale may be inversely proportional to its thickness. At room temperature the rate falls off more abruptly.* Recent measurements[30] show that a thickness of 16 Å is reached after a day and 35 Å after a year.† The latter thickness is insufficient to produce the interference colours seen on iron which has been heated for a short time in air within the range 250–350°C. Films formed on iron by exposure to unpolluted dust-free air at room temperature generally produce no change in appearance.

In contrast, the rate of *rusting* sometimes remains almost constant over considerable periods. The reason for the difference is that in scaling the oxide formed is continuous with the metallic surface and provides some protection; in rusting it may, however, happen that oxygen is reduced at one place, iron passes into solution at a second place, whilst the iron oxide appears (usually in hydrated form) at a third place, where it cannot interfere with continued attack. The contrast between scaling and rusting is

*The difference in the growth-law obeyed at high and low temperatures does not mean that the mechanism is different. Simple calculations, based on the fact that energy is needed if a particle is to move from one site on a crystal-lattice to the next, but that the amount needed can be either decreased or increased if a potential gradient exists (which will help or hinder movement according to the direction), lead to an equation of the form

$$dy/dt = A(e^{K/y} - e^{-K/y})$$

where y is the thickness of the scale at time t, whilst A and K are constant at any given temperature. (The two exponential terms represent the movement in the two directions, where the gradient helps or hinders respectively.) Under conditions prevailing at *high* temperatures, the later terms of the expansion of the two exponentials can be neglected, and the equation becomes (approximately)

$$dy/dt = K'/y \quad \text{or} \quad \tfrac{1}{2}y^2 = K' t + K''$$

where K' and K" are fresh constants. This is the so-called 'parabolic law', which is obeyed by nearly all metals over a fairly wide range of temperatures. Within the range of obedience to the parabolic law, doubling the scale-thickness merely halves the oxidation-rate; in time, thicknesses are reached at which the scale cracks off, and oxidation locally becomes rapid again (a thick film is more likely to break under stress than a thin one, for reasons easily explained). The simplest way to test whether experimental results obey the law is to plot the square of the film-thickness against the time. A straight line indicates conformity to the parabolic law. At *low* temperatures, the second exponential can be neglected entirely, but the later terms of the expansion of the first exponential now become important. Hence a different law (the 'inverse logarithmic law') is obeyed, leading to drastic slowing down of the oxidation-rate—so that the 'film' (which does not in this case deserve the name of 'scale') remains too thin even to produce interference tints; it is invisible whilst in contact with the metal, but becomes visible when separated from the basis. We can test for obedience to this 'inverse logarithmic law' by plotting the reciprocal of the thickness against the logarithm of time, measured from an appropriate zero-hour. Derivation of the various equations (involving no advanced mathematics) will be found in ref. (7) pp. 823–6 and ref. (8) pp. 298–310.

† Å stands for an Ångström Unit. 1 Å is 10^{-10} m.

seen (Fig. 1.1) on comparing (A) the Oxidation of steel plates exposed to dry oxygen at 175° and 225° with (B) the Corrosion of such plates partly immersed in salt solutions. In (A), the oxidation-rate falls off with time; in (B), after initial irregularities, the corrosion-velocity attains a constant value.

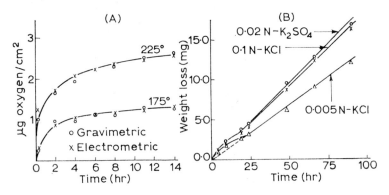

FIG. 1.1 The contrast between 'Scaling' and 'Rusting'. (A) shows the gain-in-weight plotted against time when steel is exposed at 175°C and 225°C to dry oxygen. The change starts quickly but slows down. (Davies, D. E., Evans, U. R. and Agar, J. N., *Proc. Roy. Soc.* A 1954, 225, 443.) (B) shows the loss-of-weight of steel plates partly immersed in salt solution; after preliminary fluctuation, the rate of change becomes constant. (Evans, U. R. and Hoar, T. P., *Proc. Roy. Soc.* A 1932, 137, 343.)

1.4 Conditions favourable to rusting

A horizontal iron plate fully immersed in pure water with air above the water surface slowly produces rust. Most salts, added to the water, increase the rate of rusting; but a few inhibit Corrosion almost completely—a matter to be considered later. Rusting occurs more rapidly if the plate (placed vertically or sloping) is only partly immersed, oxygen being readily replenished where the plate cuts the water-line; as explained later, there is often an unattacked zone at or just below the meniscus.

The rust formed on atmospheric exposure is usually more adherent than that produced by immersion. Exposure outdoors (when the surface is alternately wet and dry) soon sets up rusting, but in a highly polluted atmosphere sheltered surfaces sometimes rust more rapidly than those kept clean by rain. Indoors, an unpainted iron surface often remains bright for some time, except where particles of saline dust have settled on it; but if the air contains appreciable amounts of sulphur compounds derived from fuel, rusting may occur even in the absence of dust. The causes are discussed on pp. 18–26.

1.5 Electrochemical mechanism

The Corrosion of iron (or zinc) immersed in water or salt solution is undoubtedly electrochemical; the electric currents have been measured by two independent methods (ref. (7), pp. 861–8) and correlated with the corrosion-rate in the sense of Faraday's law; the good agreement between observed and calculated rates provides quantitative evidence for the electrochemical mechanism. For atmospheric rusting the evidence is less direct, but it is difficult to explain the facts except by assuming an electrochemical mechanism. The matter is discussed further on pp. 20–22.

2 Immersed conditions

2.1 Rusting of iron partly immersed in a potassium or sodium salt solution

As shown in Fig. 1.1 (B) the corrosion-rate of iron in potassium chloride solution soon becomes constant. It depends only slightly on the solution used. A twenty-fold reduction in the concentration of potassium chloride produces only a small reduction in the corrosion-rate, nor is there much change when sulphate is substituted for chloride. Other measurements have shown that the corrosion-rate on fairly pure electrolytic iron is of the same order of magnitude as the rate obtained on steel.

The fact that the corrosion-rate is so little influenced by either metallic or liquid phase is due to the fact that it is largely controlled by happenings at the meniscus, the only region where oxygen can readily be replenished from the air. Here oxygen is reduced by a cathodic reaction,* which occurs in steps, but can be summarized:

$$O_2 + 2H_2O + 4e = 4\,OH^-$$

The reaction requires four electrons, which are supplied by iron entering the liquid as cations at points lower down the specimen by an anodic re-reaction; this also occurs in steps, but can be summarized:

* If a 'plating cell' consisting of two vertical copper plates immersed in a solution of copper sulphate ($CuSO_4$, dissociating into Cu^{2+} cations and SO_4^{2-} anions) is joined to an accumulator, Cu^{2+} cations, being positively charged, will migrate to the negative plate, that is the plate joined to the negative pole of the accumulator. This plate is called the *cathode* because it is the pole to which the cations move. On arrival, each cation receives two electrons from the accumulator and thus becomes an atom. Consequently, metallic copper is deposited on the cathode; the cathodic reaction can be written $Cu^{2+} + 2e = Cu$. The other plate is called the *anode*, because it is the pole towards which the anions (SO_4^{2-}) move. Here copper enters the liquid, which is thus kept replenished. The anodic reaction is the opposite of the cathodic reaction, and can be written $Cu = Cu^{2+} + 2e$ (NOTE: the Cu^{2+} ions are hydrated, so that some Authorities like to write them Cu^{2+} (aq)).

If an ammeter is connected between the plating cell and the accumulator, the terminal marked + should be joined to the cathode of the plating cell, and that marked − to the negative terminal of the accumulator. Some confusion is often experienced, arising from the fact that for a *current-consuming cell* the cathode is the *negative* pole, since a negative charge is needed to attract the positively charged cations, whereas, in a *current-producing* cell, the cathode is the *positive* pole; for instance, if a Daniell battery is being used to provide current, the cathode, to which Cu^{2+} ions are moving, must be joined to the + terminal of the ammeter, because the positive ions arriving there are being furnished with electrons and thus keeping the copper electrode positively charged.

$$2Fe = 2Fe^{2+} + 4e$$

Since the main ions in a potassium chloride solution are K^+ and Cl^-, the cathodic and anodic products can be regarded as potassium hydroxide and ferrous chloride*; where they meet they will interact to give a precipitate of yellow-brown rust consisting of hydrated ferric oxide, Fe_2O_3, H_2O or FeO.OH—provided that plenty of oxygen is present; if a narrow vessel is used, restricting the oxygen-supply, a ferroso-ferric compound may be formed. All these facts have been established by direct observation. The alkali and ferrous salt can be detected by simple chemical tests; the precipitation and settlement of the rust are obvious to the eye. The electric current, representing the upward movement of electrons through the iron from anodic to cathodic zone, can be measured directly on a meter if the specimen is cut along the line dividing the expected anodic area from the expected cathodic area (Fig.2.1(a)); this is, however, a less accurate method of measuring the total current flowing than certain other methods available (ref. (7), pp. 861–5, 879).

The 'corrosion pattern' varies with the physical character of the metal surface. On carefully rolled steel the anodic attack may occur only along the cut edges (Figure 2.1(b)), where the internal stresses left by the shearing prevent the maintenance of a protective film. With less carefully prepared steel, corrosion starts at surface defects on the face and spreads

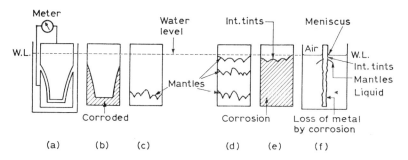

FIG. 2.1. The corrosion pattern obtained on a vertical iron plate partly immersed in potassium chloride solution. (a) shows, diagrammatically, the method of detecting the electric current flowing between the cathodic and anodic areas. (b) is the distribution obtained on high-quality metal, carefully prepared. (c), (d) and (e) refer to metal of poorer quality of careless surface preparation, and represent (c) the early stages of attack (d) later stages and (e) the final stage; (f) represents the final stage seen from the side. (Evans, U. R., *Quart. Rev.* 1967, XXI, 1.)

* The terms ferric, ferrous and ferroso-ferric which are commonly used in works on corrosion science, mean the same as iron(III), iron(II) and iron with an oxidation number between 2 and 3.

downwards and sideways over arch-shaped areas. These are first seen at points low down (Fig. 2.1(c)) and later at points higher up (Fig. 2.1(d)). Arch-shaped 'mantles' of membranous rust grow out roughly at right-angles to the surface, representing the surfaces separating regions wherein OH^- and Fe^{2+} respectively predominate. In the end (Fig. 2.1(e) and (f)), nearly the whole surface is corroding, except the zone close to the water-line, where any iron ions escaping from the metal would find themselves in alkaline liquid, so that precipitation must occur in contact with the metal; this area suffers no appreciable corrosion, but after some days it may become covered with a film thick enough to display interference colours when the specimen is taken out, washed, and dried.

The main body of the rust is precipitated well away from the metal. Its composition and appearance depends on the cross-section of the containing vessel, since this controls the oxygen supply. If oxygen is plentiful, the Fe^{2+}, OH^- and O_2 may interact to give hydrated ferric oxide as the first solid phase; but it is generally assumed that ferrous hydroxide or a basic ferrous salt is first precipitated and then becomes oxidized. Such oxidation has been studied[34]. The product can be α- or γ-FeO.OH, according as the pH value is high or low. With deficiency of oxygen, green ferroso-ferric compounds may appear; these appear to be not simply hydroxides, but to contain Cl^- or SO_4^{2-}. In some geometric situations, black magnetite is formed. Since, however, a solid formed out of contact with the metallic surface cannot protect, the composition and structure of the rust is not of great importance. The rate of attack will soon become almost constant—as shown in Fig. 1.1(b)—unless the liquid is one capable of forming a sparingly soluble substance as cathodic or anodic product. Such liquids are said to be *inhibitive* and produce little or no attack; the two cases are known as *cathodic* and *anodic inhibition*. Substances which prevent corrosion are called *inhibitors*. Inhibition is discussed on pp. 9–14.

2.2 Drop corrosion

Some early quantitative experiments[28] on drops of salt solution placed upon a horizontal steel surface, published in 1924, are worth recalling. Anodic attack occurred at the centre, the part least accessible to oxygen, and alkali was produced round the periphery. Where the alkali met the iron salt formed at the centre by anodic attack, a ring of rust appeared. After a time the drop started to spread over the surrounding (originally dry) region, producing a moist patch, within which there were discrete droplets much smaller than the original drop. The phenomenon is difficult to explain except by supposing that the cathodic reaction occurs preferentially on certain favourable spots, where the alkali produced, being hygroscopic, is capable of absorbing moisture from the air to form the droplets. Other evidence of the localization of the cathodic reaction is available today. Drop corrosion is discussed on p. 23, and is illustrated on plate 2 placed opposite that page.

Importance of solubility of the immediate products.

The mechanism of rusting sketched above is applicable to salts like potassium (or sodium) chloride (or sulphate), where both cathodic and anodic products are freely soluble. When either the anodic product or the cathodic product is sparingly soluble, the attack may be stifled or at least retarded. These two cases (anodic and cathodic inhibition) must be considered in turn.

2.3   Anodic inhibition

The criterion as to whether an anodic reaction will produce a soluble salt or a protective oxide-film was discussed at length in a lecture delivered to the First International Corrosion Conference in 1961 (ref. (8), p. 353). Earlier, the situation had been stated concisely in a paper on aluminium with Edeleanu[27] in the following words:

'At rates of attack low enough to be treated as reversible changes, if the solution is unsaturated with respect to solid hydroxide or oxide (whichever is the stable phase), the anodic reaction will lead to a soluble salt; if it is already saturated, a solid film will be formed on the anode.'

A good example of this principle was provided by Lorking and Mayne[33], who showed that certain salt solutions (such as 0.1M sodium chloride) which normally attack aluminium fail to do so if they have been saturated with oxide by several weeks' contact with alumina.

We may apply the same principle to iron immersed in dilute sodium hydroxide; all the oxides of iron are sparingly soluble in dilute alkali, and thus, so soon as the film of liquid next to the metal has become saturated with oxide, the expected reaction is the formation of a protective film. This accords with observation. Iron placed in dilute alkali undergoes no visible change, and electron diffraction studies[35] show that in fact a thin invisible film of cubic oxide is present; there is no visible rust.

Surprise is sometimes expressed that the film is oxide; hydroxide—it is argued—might be expected in a liquid so full of OH^-. However, it is easy to picture the formation of anhydrous oxide. If, at certain points favourable to the cathodic reaction, oxygen is reduced, giving further OH^-, this must be balanced by anodic reaction elsewhere to provide the electrons. The most probably reaction at the anodic points is that indicated in Fig. 2.2(a), representing the commencement of an oxide-film—followed by Fig. 2.2(b), representing further thickening. At each stage two iron atoms associate themselves as cations with the oxygen parts of three OH^- groups, allowing the hydrogen parts to join with three other OH^- groups forming water. This leads to a film of ferric oxide, Fe_2O_3. The electricity transfer is the same as would occur if the two iron atoms entered the solution as $2Fe^{3+}$, but the reaction will take place more easily; it produces immediately a situation of low energy (i.e. high stability), and avoids the passage of positive particles through a positive zone—which would have to occur if the iron passed out into the liquid. Oxide-formation is the reaction which considerations of energy would predict, and it is, in fact, observed.

The invisible protective films formed on iron by anodic inhibitors (generally cubic oxide in crystalline continuity with the metal) should be sharply distinguished from visible non-protective rust, consisting of compounds precipitated by secondary reactions; the latter are generally not cubic, but their non-protective character should be attributed more to the place of formation than to an unfavourable crystal-structure. Cubic material, if formed by secondary reactions, can be non-protective. Early work at the Massachusetts Institute of Technology[29] showed that, under conditions of poor oxygen supply, a layer of black granular magnetite (a cubic compound) may be formed which has no protective properties. It is probably formed by interaction between ferrous and ferric hydroxides; if so, it will not be in crystalline continuity with the metal—explaining the absence of protective properties.*

Alkali prevents corrosion completely if salts are absent. If sufficient chloride exists in the liquid, attack occurs. Apparently, at places where Cl^- ions carpet the iron instead of OH^-, the Fe^{2+} passes right out into the solution, without hindrance (Fig. 2.2(c)). The attack is wide-spread if Cl^- ions are in great excess of OH^- ions, but localized if it is present in small amounts; this localized attack can be very intense (Fig. 2.3 (d)), owing to the dangerous combination of large cathodic area and small anodic area, discussed on p. 32; a chloride solution containing alkali in amount just insufficient to prevent attack can produce perforation of steel sheet within a time at which, with the same concentration of chloride but without alkali, there would be only harmless general attack. Thus anodic inhibitors can be dangerous. Cathodic inhibitors, although they may not repress corrosion completely, are less dangerous when added in insufficient quantity. They will now be discussed.

2.4 Cathodic inhibition

Certain salts, notably those of magnesium, calcium, and zinc, produce adherent solid by the cathodic reaction, which interacts with iron compounds formed by the anodic reaction to give a clinging type of rust, possessing some protective properties.

Some early experiments on steel rectangles partly immersed in magnesium sulphate solution for 26 days are worth recalling (ref. (7), p. 94). In all cases magnesium sulphate solution caused definitely less corrosion than the distilled water from which the solution had been made; thus mag-

---

* The protective films on 'passive' iron generally consist of a cubic oxide, being either magnetite (Fe_3O_4), gamma-ferric oxide (γ-Fe_2O_3) or perhaps a solid solution of intermediate composition. Non-protective films are generally non-cubic; this has given rise to the belief that cubic packing confers special protective properties— which is probably wrong. In general, freedom from pores and lattice-defects is likely to be more important in determining protective quality than the system in which the atoms are packed in an ideal (defect-free) crystal.

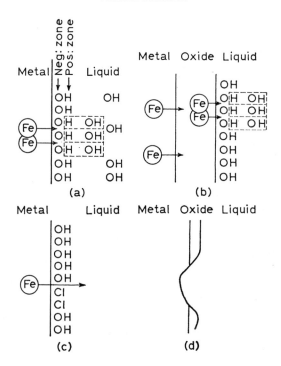

FIG. 2.2 Mechanism of the formation of an invisible, protective oxide film on iron immersed in sodium hydroxide solution; (a) early stages (b) later stages (c), (d) local attack in solution containing sodium chloride as well as sodium hydroxide. (Evans, U. R., *Corrosion Science*, 1969, 9, 816.)

nesium sulphate must be regarded as an inhibitor. Concentrations of magnesium sulphate between 0.10 M and 0.45 M produced, within experimental error, the same loss of weight in a given time.

Although over long periods magnesium sulphate has the opposite effect to potassium sulphate (diminishing corrosion instead of increasing it), the observations in the first few hours are closely analogous. In both cases corrosion starts at scattered points and spreads out downwards and sideways, producing arch-shaped areas of attack. At first the starting-points are situated mainly on the lower part of the specimen, where oxygen will be less readily replenished; ultimately the whole immersed area is suffering attack, except for a protected area close to the meniscus, which, in the case of magnesium sulphate, is a narrow strip parallel to the meniscus and perhaps 2 mm broad; in potassium sulphate, the protected zone is larger and more irregular. The main difference between the two salts is that, whereas in potassium sulphate the cathodic product (potassium hydroxide) is freely

B

soluble and diffuses away, the corresponding hydroxide of magnesium is sparingly soluble and remains as a deposit, which is first white, but later becomes bright green and ultimately pale brown; only the narrow protected zone at the meniscus remains white (Fig. 2.3).

These observations can be explained if—as suggested above—the cathodic reaction occurs only at a limited number of favourable points. After a point has become covered with magnesium hydroxide, it contributes little or nothing further to the cathodic reaction; provided the concentration of magnesium salt is sufficient to produce the necessary covering, the exact value of the concentration does not matter—explaining the fact that corrosion-velocity can be independent of concentration. But the covering up of favourable points need not prevent anodic attack on the surrounding area, and the ferrous salts formed will convert the magnesia first to green ferrosoferric compounds and then, with the gradual arrival of more oxygen, to a pale brown ferric deposit.

The exact character of the bright green and pale brown layers deserves further study. It has long been known that, when iron is placed in a sodium or potassium salt solution with restricted oxygen supply, dark green material is precipitated where, with plentiful oxygen, we should obtain

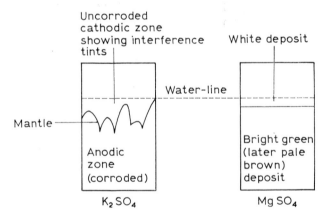

FIG. 2.3 Corrosion patterns on vertical steel plates partly immersed in potassium sulphate and magnesium sulphate solutions. In both salts, the attack starts at certain points and extends over arch-shaped areas. As shown in Fig. 2.1 (c) (d) and (e) on page 7. But in the case of magnesium sulphate, the surface just above the arch-shaped areas becomes covered with the white cathodic product (magnesium hydroxide) in the final stage, the boundary between anodic and cathodic regions remains irregular in potassium sulphate solution. It becomes straight in magnesium sulphate, where the meniscus zone is covered with an adherent white deposit.

brown rust; it is conveniently called hydrated magnetite, although, since SO_4^{2-}, Cl^- or some other anion is usually present, it would more accurately be described as a ferroso-ferric basic salt. The green product formed in a magnesium salt solution is much brighter and contains magnesium; it may be a ferroso-ferric compound in which Fe^{2+} is partly replaced by Mg^{2+}. The final product, which contains magnesium and is a lighter brown than ordinary rust, is also probably a replacement product.*

Magnesium salts are rarely added in service to inhibit corrosion. In heating-plants, indeed, their presence is regarded with disfavour; if Cl^- is also present, the hydrolysis of $MgCl_2$ produces HCl vapour, leading to attack on the relatively dry parts of the system.

The cathodic inhibitor generally used in practice is calcium bicarbonate, which is already present in many natural waters. If these also contain considerable amounts of carbonic acid, they can be quite corrosive. It may be necessary to adjust the composition by adding alkali (e.g. lime) in such quantity that the smallest *further* rise in pH—such as would arise from the cathodic reduction of oxygen at a favoured point, giving additional OH^-— will deposit calcium carbonate; a water thus treated should be non-corrosive, since the cathodic reaction is largely prevented and therefore the anodic reaction cannot proceed. The calcium carbonate deposited seems to act in the same way as magnesium hydroxide, and is later transformed to an iron compound; we get a deposit of what can be called rusty chalk or chalky rust. The situation is, however, not simple. Traces of certain organic compounds in the water—sometimes arising from pollution—can alter the physical character of the deposit; of two waters containing similar amounts of inorganic substances, one may deposit a 'nodular scale', whilst the other (containing an organic compound) may deposit an 'egg-shell scale'—which is more effectual in arresting corrosion. Information about the type of calcium carbonate scale produced is based largely on research into the behaviour of non-ferrous metals[39], but the results are helpful in connection with the rusting of iron. It was once believed that the rusty chalk prevents corrosion by providing an insulating sheath over the whole surface; conductivity measurements seem to disprove that explanation. A sheath should hinder the passage of electric current between metal and liquid, and this hindrance is not observed experimentally. If, however, the cathodic reaction occurs at isolated points and the function of the chalk is merely to put such points out of action, there is no reason to expect that conductivity measurements, carried out by conventional procedure, will be seriously affected by the deposition of chalk.

It is possible that a blob of calcium carbonate or magnesium hydroxide may merely serve as a mechanical barrier which prevents replenishment of

---

* Haematite can be regarded as a hexagonal close-packed oxygen lattice with two-thirds of its octahedral holes occupied by Fe atoms. It is possible to replace Fe by Ti and/or Mg; $MgTiO_3$ and $FeTiO_3$ are isomorphous with Fe_2O_3; $FeTiO_3$ should *not* be described as a titanate. See A. F. Wells *Structural Inorganic Chemistry* 1967, p. 486 (Clarendon Press).

oxygen at a point favourable to its cathodic reduction. Alternatively, it may serve as negative catalyst or poison for one of the steps in the cathodic reduction process.

However that may be, the deposition of chalky rust (on the inside of a water pipe, for instance) does greatly slow down corrosion. Measurements at Teddington[14] indicate that the weight-change at time t can be expressed as $\dfrac{k_1 t}{1 + k_2 t}$ where k_1 is the rate-constant, whilst k_2 introduces the blockage of sites. Clearly, as t approaches infinity, the corrosion asymptotically approaches k_1/k_2—which represents the maximum attack theoretically possible.

The performance of calcium bicarbonate already present in hard water can be greatly enhanced by the intentional addition of a second inhibitor. Much interest is being paid today to the excellent results obtained by a combination of inhibitors.

2.5 Rusting in sea-water

Magnesium and calcium are invariably found in sea-water. At one time certain experimenters, carrying out laboratory corrosion tests intended to predict the behaviour of different materials in the sea, used 0.1M sodium chloride as a substitute for sea-water. Such a practice can furnish misleading results. Generally, 0.1M sodium chloride corrodes iron more rapidly than real sea-water, which contains magnesium and calcium salts. At first the difference may not be great, but it becomes important after periods sufficient for clinging rust to be formed; in experiments lasting 128 days, sea-water, collected from the English Channel, caused only one third of the corrosion suffered in 0.1M sodium chloride (ref. (7), p. 165). In other circumstances, however, sea-water may be more corrosive than sodium chloride solution. Harbour water containing organic sulphur compounds, such as cystine,* can be very dangerous, whilst the mud of estuaries may contain sulphate-reducing bacteria, which render the oxygen of SO_4^{2-} available for the cathodic process. Atmospheric rusting of iron near the sea-shore is probably favoured by magnesium chloride in the salt, which prevents complete drying.

2.6 Rusting in pure water

Salt is not necessary for rusting. Iron or steel immersed in pure water with air or oxygen above the surface slowly develops rust. It is likely that diffusing oxygen, arriving at the metal surface in limited amounts, produces ferrous hydroxide, which is appreciably soluble; it diffuses outwards to a place where there is more oxygen, so that the less soluble hydrated ferric oxide (Fe_2O_3, H_2O or $FeO.OH$) is precipitated; this is mainly pro-

---

* *Cystine*, an organic sulphide, can be written Cy_2S_2, where Cy stands for $-CH_2 \cdot CH\,(NH_2) \cdot COOH$. Its cathodic reduction leads to *Cysteine* CyHS.

duced out of contact with the metal and is therefore non-protective. If, however, measures are adopted to ensure that oxygen can arrive in abundance at the metal surface, it is possible to produce ferric compounds on the metallic surface, and these may have some protective value.

Early experiments (ref. (6) p. 108) carried out with an 'eccentric whirler' (Fig. 2.4) demonstrated this effect, but different results were obtained according as pure iron or steel was being studied. The surface of pure iron remained bright and silvery after whirling in distilled water exposed to air; there was no change apart from the formation of tiny dark spots, probably representing the sites of pores or inclusions—which photo-micrographs

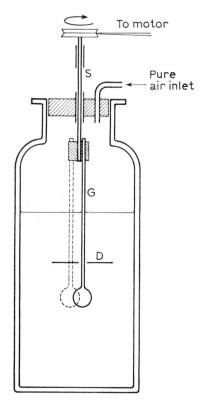

FIG. 2.4 Eccentric whirler. The metal disc D has a central hole which allows it to fit loosely on the vertical glass rod G, attached eccentrically to the shaft S. When the latter is driven by means of an external motor, the disc mounts up the rod G, and, since the point of contact between metal and glass is constantly changing, receives everywhere an ample replenishment of oxygen.

(Evans, U. R., *Corrosion of Metals*, 1926, Edward Arnold, London.)

showed to be present. (Possibly still purer iron would suffer no change at all, becoming covered with an invisible film of ferric oxide; further work, using the superior materials now available and electron diffraction to identify the protective film, would be interesting.) In contrast, mild steel developed a brassy appearance—due to microscopic blobs of a rust-like substance adhering to the surface, but evidently not constituting a continuous, protective covering. On the heterogeneous surface, the two stages (formation of dissolved ferrous hydroxide and its oxidation to the ferric state) may have occurred at different points—explaining the non-uniform and imperfectly protective character.

3 Atmospheric conditions

3.1 Mechanism

At one time atmospheric rusting was regarded as a form of dry corrosion —comparable to direct oxidation. It is now believed to be more akin to drop corrosion. The important work of Vernon[22], Bukowiecki[24] and others makes it clear that rusting only takes place at an appreciable rate under conditions of humidity consistent with the presence of a liquid phase, even though the trace of liquid causing the damage cannot generally be seen.

3.2 Rusting by salt dust

The rusting set up by the settlement of saline particles can only occur if the humidity of the air is sufficiently high to render stable a liquid film; a visible pool of liquid is not needed. A saturated solution of sodium bromide stands in equilibrium with air of 59% R.H.* at 20°C; thus a solid particle of sodium bromide will become damp if the air humidity exceeds 59%. It was found by Bukowiecki that a steel surface carrying particles of sodium bromide remains unrusted when exposed to air of 50% R.H. but develops rust at 60% R.H. The passage between immunity and rusting is not always sharp. Saturated sodium chloride is in equilibrium with air of 78% R.H. In Bukowiecki's experiments, steel carrying sodium chloride particles became strongly rusted at 80% R.H., and remained bright at 60%; at 70% the salt particles became brown and there was perceptible attack on the metal. Strongly hygroscopic salts can set up rusting even in relatively dry air; examples are provided by the chlorides of zinc, lithium, magnesium and calcium; their saturated solutions stand in equilibrium with air at 10%, 15%, 32% and 32.3% R.H. respectively.

If it is recognized that rusting produced by a solution of a salt is electrochemical, it would seem that the rust produced by saline particles must also have an electrochemical mechanism.

The rusting set up at a point where a salt-particle rests on iron or steel may spread over the metal surface, if humidity conditions are favourable. Uniform spreading gives expanding circles of rust. If, through some

---

* R. H. stands for 'relative humidity', the moisture-content of the air expressed as a percentage of the moisture-content of 'saturated' air—that is, air in equilibrium with a *flat* surface of water.

irregularity, advance commences more rapidly at certain points on a periphery than others, thread-like out-growths of rust are formed. This is known as *filiform* corrosion. A study of the phenomenon at Teddington[37] seems to favour an electrochemical mechanism.

3.3   Rusting in moist air containing sulphur dioxide

The atmospheric corrosion set up in industrial or urban atmospheres containing sulphur dioxide (arising from the combustion of coal or oil containing sulphur) presents some different features. Here there are no hygroscopic solid particles, but sulphuric acid, formed by the oxidation of sulphur dioxide, soon becomes available to collect moisture—even from

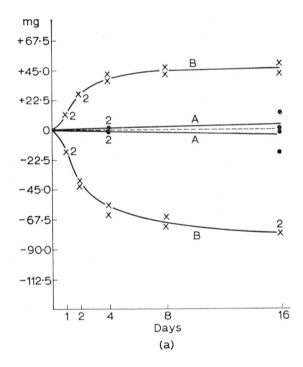

FIG. 3.1(a) Weight-changes on iron exposed to moist air with sulphur dioxide. The upper curves show gain-in-weight and the lower curves show loss-of-weight, both expressed as milligrammes for a total area of 9·6 cm². The curves show that SO_2 is needed for rusting, but only as a source of $FeSO_4$. Moist air (95% R.H.) without SO_2 causes only a small change (Curves A). If SO_2 is present as well as moisture rusting is much more rapid (Curves B).

air which is definitely unsaturated. The account of the changes presented below is based on recent laboratory experiments started at Cambridge about 1965 by the Author and later continued by C. A. J. Taylor, whose work is still unfinished. It will make for clarity to follow the line of the

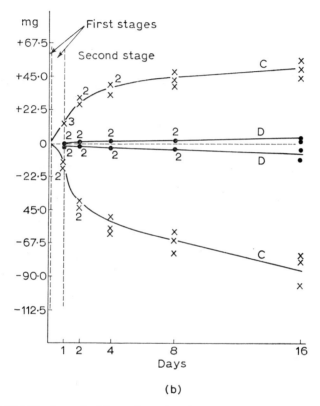

(b)

FIG. 3.1(b) If, after only 4 hours in moist air containing SO_2, the specimen is moved to moist air free from SO_2, rusting still continues rapidly (Curves C). If, between the two stages, the specimen is deeply immersed in water, so as to remove $FeSO_4$, subsequent changes are slow (Curves D). (Evans, U.R., *Corrosion Science*, 1969, 9, 818–19.)

Cambridge researches—without claiming originality for all the points observed, several of which had been recorded much earlier by Vernon[22], Schikorr[17] and other investigators.

In the experiments of 1965, specimens of iron were exposed in closed glass tubes to humid air containing sulphur dioxide; the humidity was regulated by a small amount of sulphuric acid of known concentration placed at the bottom of each tube, but not reaching the specimen. The first change was the formation of a film of moisture, barely visible to the eye; if, at this stage, one of the tubes was opened and the iron surface wiped with filter-paper, tests showed that ferrous sulphate had been formed; it was detected before any solid corrosion-product could be seen.

Later, grey specks appeared and these gradually became brown; that was the first appearance of ferric rust (Fe_2O_3, H_2O, generally written FeO.OH). When once ferric rust and ferrous sulphate were present, it was found possible to move a specimen to another tube containing moist air free from sulphur dioxide, and yet it was observed that rusting continued. It would seem that sulphur dioxide is needed to start rapid atmospheric rusting, but serves only as the source of ferrous sulphate; when once ferrous sulphate has been formed, sulphur dioxide is no longer needed.

The situation was examined quantitatively by Taylor, some of whose results are reproduced in Fig. 3.1. Each set of experiments provides two curves—the upper one showing the *gain-in-weight* observed after different times of rusting, and the lower one the *loss-of-weight* after de-rusting. If the rust consisted of FeO.OH and no other component, the loss should be 1.69 times the gain—this being the value of Fe/OOH; there is approximation to that ratio for the longer experiments, but not for the shorter ones, where other compounds such as hydrated ferrous sulphate still represent an appreciable part of the rust. The curves of Fig. 3.1 show the importance of sulphur compounds. Moist air of R.H. 85%, without sulphur dioxide, produces only a small change (Curves A). If sulphur dioxide is present in air of the same humidity, changes are much more rapid (Curves B). If, after the specimen has stood 4 hours in moist air containing sulphur dioxide, the specimen is moved to moist air free from sulphur dioxide, the results (Curves C) are hardly different from those obtained when the specimen is left in the original tube (Curves B). If, after removal from the tube containing sulphur dioxide, the specimen is placed in water for 24 hours and then exposed in moist air free from sulphur dioxide, the change is slow (Curves D); for the water has dissolved out all soluble compounds, including ferrous sulphate and sulphuric acid.

An interesting point is that each molecule of ferrous sulphate can produce *many* molecules of rust. Any mechanism proposed for the atmospheric rusting of iron must explain this, if it is to gain acceptance. The following seems to satisfy requirements.

At the outset an adsorbed moisture film on the metal surface dissolves sulphur dioxide, which is oxidized by air to sulphuric acid. This attacks the iron giving ferrous sulphate, and part of the latter is oxidized by air to ferric sulphate and ferric hydroxide (rust)*

---

* Probably all these reactions follow an electrochemical course. The oxygen is reduced at points suitable to the cathodic reaction

$$\tfrac{1}{2}O_2 + 2H_3O^+ + 2e = 3H_2O$$

So soon as ferric rust containing ferrous sulphate has appeared on the iron surface, conditions are suitable for an *Electrochemical Cycle*, which will produce additional ferric rust and regenerate ferrous sulphate. Anodic attack upon the iron can be written:

$$Fe = Fe^{2+} + 2e$$

and this will be balanced by the cathodic reduction of brown rust to black magnetite,

$$8\ FeOOH + Fe^{2+} + 2e = 3Fe_3O_4 + 4H_2O$$

The electrons provided by the anodic reaction are used up in the cathodic reaction. The magnetite will then be re-oxidized by air to ferric rust, which can be written:

$$3\ Fe_3O_4 + 0.75\ O_2 + 4.5\ H_2O = 9\ FeO.OH$$

It will be noticed that the amount of rust has increased; 8 FeO.OH has become 9 FeO.OH. One Fe^{2+} produced by anodic destruction of the iron is balanced by one Fe^{2+} removed into the rust-phase by the cathodic reaction. Since SO_4^{2-} is not destroyed in the cycle as presented, it follows that the total of $FeSO_4$ remains unchanged—explaining why one molecule of $FeSO_4$ can produce many molecules of rust. Indeed the equations as printed above would suggest that a finite amount of $FeSO_4$ could produce an *infinite* amount of rust. The curves of Fig. 3.1, however, show a flattening, which indicates that a given amount of ferrous sulphate can only produce a *finite* amount of rusting. This is explained by the fact that when ferrous sulphate solution is exposed to air, the oxidation and hydrolysis

---

thus maintaining anodic oxidation of SO_3^{2-} (the anion of H_2SO_3, produced from SO_2 and H_2O), of Fe and of Fe^{2+}

$$SO_3^{2-} + 3H_2O = SO_4^{2-} + 2H_3O^+ + 2e$$
$$Fe = Fe^{2+} + 2e$$
$$Fe^{2+} = Fe^{3+} + e$$

The electrochemical mechanism, however, is not universally accepted. It is sometimes stated that rust catalyses the oxidation of SO_2 to H_2SO_4, but it should be noted that ferrous sulphate is formed before any rust appears. Some chemists would probably write the oxidation of ferrous sulphate to ferric sulphate and ferric hydroxide thus:

$$3FeSO_4 + 0.75\ O_2 + 1.5\ H_2O = Fe_2(SO_4)_3 + Fe(OH)_3$$

The ferric sulphate may undergo hydrolysis to some extent, giving basic ferric sulphate and sulphuric acid, which may attack a further quantity of iron. It has been suggested that the continuation of rusting is due to the attack on the iron by the acid liberated by hydrolysis, producing more ferrous sulphate and thus in turn more acid. Such reactions can in fact occur, but, except in the opening stages, they probably account for only a very small part of the rusting observed. The Electrochemical Cycle operates much more rapidly than the Acid Regeneration Cycle.

leads to the formation of a basic ferric sulphate, so that SO_4^{2-} is gradually withdrawn from the cycle and the rusting slows down.

These statements are not just based on paper calculations; the reactions have been demonstrated in the laboratory. In order to study the electro-chemical reduction, a layer of iron rust was produced on a *copper base*; (if it were supported on an iron basis, much of the current produced would flow directly between iron and rust and would escape measurement). A sheet of copper was plated over a rectangular area with iron, which was later entirely converted to rust by being wetted with ferrous sulphate solution; this was allowed to evaporate, leaving a layer of tiny crystals; the specimen was then exposed to moist air for about 2 days at 25°C, and developed in that way an adherent layer of rust. The 'rusty copper' was combined with a piece of unrusted iron (Fig. 3.2) to give the cell

$$\text{Iron} \quad \big| \quad \text{FeSO}_4 \text{ solution} \quad \big| \quad \text{FeO.OH (supported on copper)}$$

The cell was joined to a milli-ammeter, and the current generated was measured. The bright iron functioned as anode and the 'rusty copper' as cathode. The current died away with time. When it had become low, the 'rusty copper' specimen was taken out; it was observed that the originally brown deposit had become black, since the ferric rust had suffered cathodic reduction to magnetite. When the specimen was exposed wet to air, the black again became brown, since the black magnetite was oxidized once more to brown ferric rust (a larger amount than before the reduction). The specimen could then be returned to the cell, and a fresh discharge of current—similar to the first—could be obtained. It was found that the cycle (cathodic reduction of rust to magnetite with generation of current, and then revivification by oxidation in air) could be repeated many times. This demonstrated the proposed mechanism.

The manner in which the reactions take place when rusty iron is exposed to damp air at constant humidity is tentatively suggested in Fig. 3.3 (the state of affairs will vary with the atmospheric conditions, and will be more complicated out of doors where there will be alternate rainy and dry periods). In a steady state, the inner part of the rust will probably consist

PLATE 1 The Delhi Pillar, erected in the Fourth Century A.D. and moved to its present site in the Twelfth Century, is still, in its upper part, free from typical rust, although a visible oxide-layer is present. Three factors may have contributed to this exceptional performance, (a) absence of appreciable amounts of sulphur compounds in the atmosphere (b) absence of appreciable amounts of sulphur in the metal (c) compressional stress left in the metal by the hammering of the iron when the pillar was originally produced. All three factors are unfavourable to the corrosion-reactions which lead to rust, whilst not interfering with the production of a protective oxide-film. (Photograph by kind permission of B. J. Stokes.)

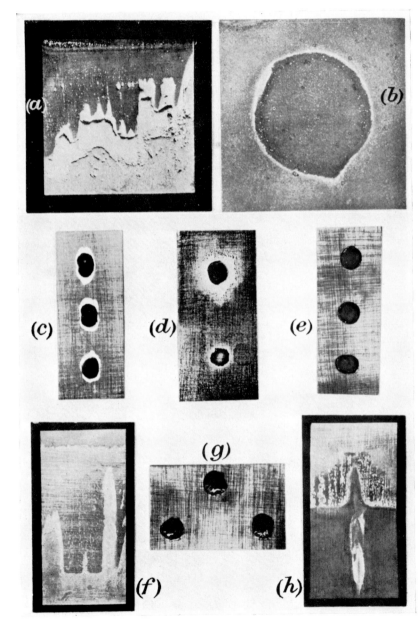

(a)

(b)

(c)

(d)

(e)

(f)

(g)

(h)

of magnetite with ferrous sulphate solution in the pores, whilst the outer part will consist of ferric rust with air in the pores. Thus iron is anodically attacked along the plane XX′ and the corresponding quantity of fresh ferric rust is formed along the plane YY′. Since metal is removed at one place and the rust deposited at another, no protective film will be formed— such as would stifle the change. This explains why atmospheric rusting, once started, will generally continue. On the other hand, the electrochemi-cal reaction can only take place if the electrical circuit is complete; this needs (1) an *electrolytic conductor* to convey the *ions* and (2) an *electronic conductor* to convey the *electrons*. The first is provided by the ferrous sul-phate solution, but if the relative humidity is too low, the solution will dry up, as explained in connection with corrosion by salt particles; below a

PLATE 2 EXPERIMENTS IN CORROSION. Exhibits (c), (d), (e) and (g) were produced by placing drops of potassium chloride solution on horizontal plates of abraded iron. Anodic attack occurs at the centre of each drop, whilst the peripheral zone, being cathodic, remains unattacked; a ring of rust appears where the anodic and cathodic products meet, and forms a loose membrane extending over the central anodic zone. (c) and (d), where the salt has been allowed to dry up, show clearly that the rust is formed only in the centre of the drop area. On (e), where most of the membrane has been removed, it can be seen that the iron within the ring has been strongly attacked. On (g), representing an experiment carried out over water in a closed vessel, so that evaporation was avoided, the wrinkling membrane of rust can be seen. The distribution of attack is an example of Differential Aeration; the outer zone of each drop, where oxygen is best replenished, is cathodic, and remains unattacked; the centre suffers anodic corrosion.

Similar effects can, under some conditions, be produced on non-ferrous metals. Exhibits (a) and (f) are plates of zinc partly immersed in potassium chloride solution. In each case the upper part, where oxygen can be re-plenished, is cathodic and remains unattacked; the corrosion occurs lower down; similar effects can be obtained on a partly immersed plate of iron, but here the corrosion-product is not white. On brass it is necessary to exclude oxygen more effectually if anodic attack is to be set up. Exhibit (h) shows a brass plate which has been leaning against a glass rod whilst immersed in a salt solution; in the crevice between the brass and glass, there is little re-plenishment of oxygen, and marked corrosion can be seen as a vertical white streak. Exhibit (b) represents a horizontal lead plate on which a glass lens has been placed and the whole immersed in potassium chloride solution. Anodic attack, producing a dark ring, has taken place in the crevice; the white points visible are due to light reflected from pits. Further out on the areas accessible to oxygen, the cathodic reaction has taken place. The white ring of lead hydroxide, precipitated where the anodic product (lead chloride) and the cathodic product (potassium hydroxide) have met, is clearly visible.
From Evans, *Corrosion of Metals* (1924), Edward Arnold.

R.H. value of about 70%, atmospheric rusting of iron in air containing SO_2 is found to become very slow.

The electronic conduction is provided by the magnetite. The existence of an intermediate oxide of high electronic conductivity explains the fact

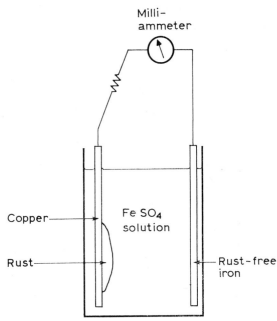

Milli–
ammeter

Copper

Fe SO_4
solution

Rust

Rust-free
iron

FIG. 3.2 Reduction of ferric rust to magnetite and re-oxidation by air. The copper carrying rust with rust-free iron as anode, are connected through a milliammeter. A current flows, but after a time dies away, and it is then found that the brown FeOOH has been converted entirely to black Fe_3O_4. On exposure wet to the air it again becomes brown, owing to re-oxidation to FeOOH, and now a fresh spell of current can be obtained.

that the atmospheric corrosion of iron proceeds more quickly than that of zinc, which under immersed conditions is often corroded more quickly than iron—a matter discussed later.

It is interesting to enquire why magnetite is a better conductor of electrons than most other oxides. Magnetite is a member of a large group of double oxides (crystallizing in the cubic system) known as *spinels*, formed by the combination of oxides of divalent and trivalent metals; the general formula is $MO.M'_2O_3$ where M is the divalent and M' the trivalent metal. Magnetite (Fe_3O_4 or $FeO.Fe_2O_3$) differs from other spinels in that M and M' both represent iron. Crystals of magnetite contain Fe^{2+} and Fe^{3+} ions in contact. At any point, an electron can pass from Fe^{2+} to Fe^{3+}, the first now becomes Fe^{3+} and the second Fe^{2+}, so that the energy situation is unchanged. This would not be true of spinels in which M and M' are different elements; for instance, if M' was Cr and M remained Fe (as in the mineral chromite), an increase in energy would be needed for an electron to pass from the Fe^{2+} to the Cr^{3+}. This may explain why steels containing small amounts of chromium suffer atmospheric attack more slowly than ordinary mild steel—as suggested later.

There is no doubt about the accelerating effect of sulphur dioxide in industrial or urban air. Hudson and Stanners[31] have made careful measurements of atmospheric corrosion in various localities and have established a definite correlation between the corrosion-rate and the sulphur-dioxide content of the atmosphere. Schikorr[38] has studied corrosion in a definite locality, but at different seasons of the year, and has found that it is most rapid when sulphur dioxide is present in largest amount.

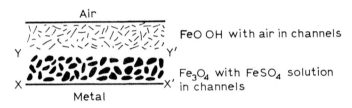

Air

FeO OH with air in channels

Fe_3O_4 with FeSO$_4$ solution in channels

Metal

FIG 3.3 Tentative suggestion for the electrochemical mechanism of atmospheric rusting. Anodic attack occurs at level XX' and cathodic reduction of ferric rust to magnetite at level YY'. The freshly formed magnetite is re-oxidized by air, so that as much iron enters the ferric phase as is lost by anodic attack from the metallic phase. (Evans, U. R., *Corrosion Science*, 1969, 9, 820.)

Mayne[34] has watched the behaviour of painted specimens, on which the paint had been applied over rust. If the painting has been carried out in winter (when the rust contains much ferrous sulphate), the paint-coat breaks down more quickly than if painting has taken place in summer, (when there is less ferrous sulphate in the rust). The ferrous sulphate generally occurs in nests or patches, and favours the formation of fresh rust below the paint; this occupies a much greater volume than the iron destroyed in producing it, so that the paint-coat is locally humped up. Its power of elastic distension is insufficient to accommodate the expansion; consequently the paint-coat breaks, leaving the iron locally unprotected. This emphasizes the need to remove all rust from iron before painting—a matter which will be discussed later. In fact, however, it is not the rust itself, but the salts present in the rust (especially the iron salts), which do the mischief. A patch of rust produced by placing a drop of distilled water on iron and allowing it to evaporate before paint is applied does not cause premature failure of the paint-coat.

The mechanism of atmospheric corrosion is still under investigation at Cambridge; readers interested in the subject may care to study a new paper by C. A. J. Taylor and the Author of this book (reference on p. v).

3.4 Corrosion probability

Common observation shows that when steel is exposed to the atmosphere rusting starts locally; parts of the surface still remain bright even after long periods. It is necessary, therefore, to introduce a new concept—the probability that corrosion will start at all within a certain definite area. One method of measuring this probability (ref. (7), p. 936) consists of placing drops of distilled water of a definite size on a horizontal surface and counting the proportion which develop rust. When the atmosphere is an oxygen-nitrogen mixture, the *probability* declines as the oxygen-content is raised, although the *conditional velocity* of the corrosion (the rate of corrosion proceeding below those drops which cause any rusting at all) increases. Below those drops which suffer no visible change, the iron has presumably developed a protective film.

Laboratory experiments have shown that the presence of sulphur in the steel, and also its presence in air (generally as sulphur dioxide), can increase corrosion probability and also conditional velocity. This explains the bad behaviour of unpainted steel exposed to air under conditions existing today in most towns and industrial regions and also the surprising good behaviour of ancient iron exposed to an atmosphere free from sulphur compounds.

3.5 Oriental iron

Iron produced in Eastern countries in early times, with charcoal as fuel, must have contained very little sulphur; some of the old iron beams, pillars, or chains have been set in situations which are still remote from industrialization, where the air has remained almost free from sulphur dioxide. The Delhi pillar, probably erected in the fourth century A.D. and moved to the present site in the twelfth century, provides an example. (See Plate 1.) A recent analysis records only a 'trace' of sulphur in the portion above ground and 0.008% in the underground part. The upper portion of the pillar has remained rust-free and the surface is described by different writers as bronzy or as bluish; evidently, of the two alternative reactions, oxide-formation has prevailed over rust-formation, and during sixteen centuries exposure at temperatures periodically elevated by the sun, the film has reached visible thickness. It seems likely that the immunity is due to the fact that sulphur is almost absent both in the metal and in the atmosphere; the probability of rust appearing is therefore extremely low. Some authors attribute the good behaviour of the iron to the absence of manganese or the presence of phosphorus. Another authority ascribes the absence of rust entirely to the unpolluted atmosphere; this is almost certainly an important cause, but it is likely that the absence of sulphur in the metal has also contributed to the remarkably good state of preservation.

3.6 Mediaeval British wrought iron

The good performance of wrought iron in our mediaeval buildings is responsible for the belief that, if only the traditional process of making wrought iron could be revived, rusting could today be avoided. It is, however, probable that the absence of sulphur compounds from the air in early days was responsible for a comparative absence of corrosion trouble. Exposure tests carried out by Hudson[10] at various places in the U.K. about 1931–1938 showed that wrought iron suffered corrosion only slightly more slowly than ordinary mild steel and at a rate similar to steel containing 0.2% of copper. It is well to note that the wrought iron used for these tests probably contained more sulphur than the iron made in mediaeval times when charcoal was used as fuel both in the puddling furnace and the forge. But in these same tests modern Swedish iron made with charcoal corroded more rapidly than the British iron. The subject is complicated by the fact that traditional British wrought iron, after puddling and piling*, contains much slag, and consists of a series of parallel layers, some very susceptible to attack and others resistant (ref. (7), pp. 508–513); as a result, the corrosion-velocity measured parallel to the layers is much faster than that measured in the direction normal to the surface. It would, however, not be possible for an artist, beating out iron into ornaments at his forge, to arrange that the resistant layers should always be kept parallel to the surface.

There is little doubt that it is the increased sulphur content of the air, and not that of the metal, which explains the enhanced rusting of modern times; a revival of mediaeval iron-making processes would not prevent rusting.

3.7 Low-alloy steels

It has been mentioned that the addition of about 0.2% of copper to mild steel reduces the corrosion-rate roughly to the level of wrought iron. Small amounts of nickel, chromium, aluminium and molybdenum (often in combination with copper) confer even greater resistance, as shown by extensive tests in the U.S.A.[13] and the U.K.[10]. These 'low-alloy' steels cannot be described as non-rusting. They must be distinguished from the 'stainless steels' containing 13% of chromium (or, in the austenitic type,

---

* In the process favoured at many iron-works, the puddled iron was rolled into plate which was then cut into rectangles. These were placed on one another, giving a *pile*, which was passed once more between the rollers; the rectangular components became welded together. The process—now almost obsolete—was known as *piling*.

c

18% chromium and 8% nickel).* They do, however, corrode more slowly than unalloyed steel, and are much cheaper than any stainless steel. One commercial product with 0.5% copper, 1.0% chromium, 0.16% phosphorus, and 0.8% silicon corrodes in the atmosphere at about one-third the rate of ordinary mild steel. The corrosion-rate of low-alloy steels tends to fall off with time. In certain parts of the World, these materials are being used today in structural work without protection by paint. There are

* To discuss stainless steel properly would require another book, but a brief review may be provided. In Victorian days, the need to clean domestic cutlery after each meal was an irksome chore, although machines with rotary brushes worked from a handle reduced the manual labour. The introduction, after the 1914–18 war, of a ferritic steel with 13 % chromium, suitable for table-knives, brought relief, but complaints were soon heard that they were blunt. With increasing experience in grinding and hardening, these complaints diminished, but quite possibly the cleaning brushes had been acting as strops, and the knives, as they reached the table, had really been sharper in the good old days. These ferritic steels were unsuited for outdoor conditions, but later austenitic steels with a higher chromium content (18 %), along with 8 % nickel to stabilize the austenitic structure, were introduced and have proved very useful—although not suitable for cutlery.

It is commonly stated that stainless steels owe their corrosion-resistance to an invisible protective oxide-film. Such a film is certainly present, but, since a similar film is present on ordinary (chromium-free) steel, this cannot provide the complete explanation. The films on the two materials, however, behave differently when exposed to mildly acid conditions, such as are usually present in a British atmosphere. Any oxide-film periodically keeps cracking, especially at places where the metallic basis carries tensional stresses, but in a dry, unpolluted atmosphere, the break will immediately repair itself by fresh oxide-formation. If, however, acid moisture is present, the damage may extend, since the oxide around the gap will pass into solution—not by direct dissolution as Fe^{3+} ions, but by 'reductive dissolution' as Fe^{2+} ions; Pryor showed in an extensive series of researches that reductive dissolution is a much more rapid change than direct dissolution. The reactions of reductive dissolution can be written thus:

Anodic dissolution of *iron* exposed *at* the gap, $Fe = Fe^{2+} + 2e$

Cathodic dissolution of *film around* the gap, $Fe_2O_3 + 6H^+ + 2e = 2Fe^{2+} + 3H_2O$

As a result, the bare area rapidly extends, the steel becomes film-free, and corrosion sets in.

If the material contains chromium, reductive dissolution would require the formation of Cr^{2+} which is much less stable than Fe^{2+}; provided that a sufficient concentration of oxygen is present, the cathodic reduction of that oxygen will take place preferentially to the reductive dissolution of the film, so that the gap will heal itself by fresh formation of oxide. The amount of oxygen needed has been studied in an extensive research by Berwick. Starting with stainless steel in the active condition, he found that, if the oxygen-content of the acid exceeded a certain amount, the steel became passive—as was shown by a rise in potential; the time needed depended on the oxygen-concentration and other conditions. In other experiments, he started with the material passive, and showed that if oxygen is excluded the steel can become active, as shown by a falling potential. The critical oxygen-content depends on the sulphuric acid concentration, but under atmospheric conditions there will always be a sufficient oxygen supply to prevent reductive dissolution, so that the film becomes, and/or remains, protective. Any slow direct dissolution will easily be made good by fresh oxidation. For details, the original paper (I. D. G. Berwick and U. R. Evans, *J. appl. Chem.* 1952, 2, 575) should be studied, and also the researches on direct and reductive dissolution (M. J. Pryor & U. R. Evans, *J. chem. Soc.* 1949, p. 3330; 1950, pp. 1259, 1266, 1274). A paper entitled 'Why does stainless steel resist acids?' (U.R. Evans, Chem. Ind. 1962, p. 1779) may be consulted. The whole subject of stainless steel is authoritatively discussed by J. H. G. Monypenny, *Stainless Iron & Steel* (Chapman & Hall).

favourable accounts of their behaviour, unpainted, on buildings in the Middle West States of America. It is unlikely that they will behave equally well in atmospheres where salt and/or sulphur dioxide are present in large amounts. It would be premature to attempt a definite judgement regarding the use of such materials, unpainted, in Europe. Tests, are, however, being carried out in European countries, and the results are awaited with interest.

Opinion is divided regarding the causes of the relatively good behaviour of low-alloy steels, but the mechanism for the rusting of unalloyed steel already proposed suggests an explanation without further assumptions. The Electrochemical Cycle would predict that a given amount of ferrous sulphate, once produced, would lead to an infinite amount of rusting unless it is removed from solution, e.g. as basic ferric sulphate. Now copper and nickel form basic sulphates which are stable and sparingly soluble. Small amounts of these metals in steel may be expected to remove SO_4^{2-} from the cycle in an insoluble form more quickly than would occur on non-alloyed steel. This is indeed confirmed by the extensive measurements of Copson[26], who showed that the amount of sulphur held in rust increases sharply with the amount of copper in the steel. He writes, 'The function of copper and nickel is to render sulphates insoluble by forming complex basic sulphates'. He seems to picture the basic sulphate as blocking the pores in the rust. Now in electrochemical action it is not always necessary for the corrosive agent to reach the seat of corrosion; it may stimulate the reaction at a cathodic point—which will in turn promote attack at an anodic point. Nevertheless a channel filled with electrolyte must connect the two points. If a sparingly soluble solid, such as basic copper sulphate, is formed in this channel, reducing its cross-section, the current flowing will be diminished and the corrosion-rate lowered. In that sense, Copson's picture of blockage appears correct, but it must be recollected that the removal of SO_4^{2-} from solution as a basic sulphate (whether of copper or nickel), will also reduce the specific conductivity and therefore diminish current-flow. Both factors can be embraced in the statement that copper and nickel reduce the corrosion-rate by increasing the resistance of the *electrolytic* limb of the circuit.*

In contrast, chromium probably acts by increasing the resistance of the *electronic* limb. As already stated, magnetite is a better conductor than most spinels because electron-exchange is possible between neighbouring Fe^{2+} and Fe^{3+} ions; the replacement of Fe^{3+} by Cr^{3+} will reduce the conductivity, and Verwey[50] has shown that chromium added to magnetite does increase the resistance.†

* Recent Japanese work indicates that the formation of protective rust on low-alloy steels containing copper and phosphorus is due to the catalytic action of these two elements in accelerating the oxidation of iron from the ferrous to ferric state; consequently a layer of amorphous FeO.OH is formed in close contact with the metal and serves to stifle attack. See Misawa, T., Kyuno, T., Suětaka, W. and Shimodaira, S., *Corr. Sci*, 1971, **11**, 35; Matsushima, I. and Veno, T. ibid., p. 129.

† Verwey added the chromium as $MgO.Cr_2O_3$, so that his observations are not directly applicable. Nevertheless, the explanation offered seems the most probable one.

Clearly, if the corrosion-product were to contain insoluble salt instead of a conducting oxide, it would be expected that the corrosion-rate would be further reduced, since most solid salts possess practically no electronic conductivity. The formation of phosphate instead of oxide should be beneficial. The additions of the metals mentioned above work best in retarding atmospheric corrosion if phosphorus also is present in the low-alloy steel.

4   Protective measures

4.1 Difference between the atmospheric behaviour of iron and non-ferrous metals

Before considering the use of non-ferrous metals as protective coatings on iron, it is pertinent to ask the reason for the relatively high resistance of most non-ferrous metals to atmospheric attack. It has been explained that the good electronic conductivity of magnetite is essential for the establishment of the electrochemical circuit needed in rapid atmospheric attack. Most non-ferrous metals form no corresponding intermediate oxide of high electronic conductivity. Thus their corrosion in the average atmosphere proceeds much more slowly than that of iron. The zinc coatings present on galvanised iron serve to protect the iron basis under many conditions, despite the fact that the E.M.F. of the cell Zinc/Oxygen is higher than that of the cell Iron/Oxygen. There are, however, conditions under which zinc is corroded more quickly than iron, and these can include atmospheric conditions, if the condensed moisture is very acid; in a railway tunnel, zinc is attacked more quickly than iron—as shown in Hudson's work. Under partly immersed conditions in (say) potassium chloride solution the attack upon zinc is found to be faster than that upon iron; the mechanism is thus analogous to that described for iron on p. 6, but the E.M.F. of the cell is higher; thus the corrosion-rate, which is equivalent to the current, is also higher.

In some situations, the rapid attack on zinc is used in providing *cathodic protection* for steel; an example is the use of zinc protector blocks attached to steel pipes buried in the ground; the zinc, being the anode of the cell zinc-iron, is sacrificed and the steel, being the cathode, receives protection. From time to time, the zinc blocks must be replaced, but this involves less inconvenience and smaller expense than the replacement of the pipe-line.

The good resistance to corrosion often shown by aluminium can be attributed to the invisible Al_2O_3 film present on the surface. This is very thin, but alumina is a bad conductor of electrons and hinders the cathodic reaction, which requires the passage of electrons between metal and liquid. Consequently, the corrosion-process is retarded. The behaviour is further improved if the alumina film is thickened by 'anodization'. When aluminium is made an anode in a suitable bath, the film thickens, and the thickness ultimately reached is roughly proportional to the E.M.F. applied.

'Anodized' aluminium is resistant under conditions where steel or zinc would be attacked.

It is, however, rash to generalize, or to claim that one material is in any absolute sense better than others; aluminium, for instance, is attacked by a dilute solution of sodium hydroxide, which has practically no action on iron. In choosing a metal for a structure (or a vessel), or in choosing a protective scheme (whether based on plating or painting), the character of the liquid or the atmosphere to which the structure (or vessel) will be exposed must be carefully considered; unless that is done, there may be destructive attack.

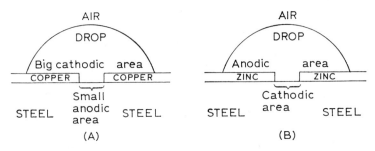

FIG. 4.1 Copper and zinc coatings on steel, showing behaviour at a pore; copper is cathodic to steel, whilst zinc is anodic. In the first case, the copper coating provides a large cathodic area on which oxygen can be reduced, providing a large current, the effect of which is concentrated on the very small area of steel exposed at the pore – so that the corrosion is intense. In the second case, the zinc is anodic to the steel, so that the latter is not attacked at all; the corrosion, distributed over a large area of zinc, does not produce serious damage.

4.2 Prevention of rusting by a coating of non-ferrous metal

An obvious method for protecting steel is to cover it with a layer of more resistant material. At first sight, a thin plating of a noble metal, which would suffer no attack, may seem the right choice; but, apart from the high cost of metals like gold, silver or platinum, the result would be disastrous if there was the smallest pin-hole in the coating. Even copper-plating, unless completely non-porous, can cause the corrosion of the steel to be locally more intense than that of unplated steel subjected to the same conditions. This is easily explained. Suppose (Fig. 4.1 (A)) that there is water or salt solution at the site of a pore. A short-circuited corrosion cell

$$\text{copper} \mid \text{liquid} \mid \text{steel}$$

will be set up. The copper will provide a relatively large cathode, providing an adequate area on which oxygen can be reduced; the whole of the anodic attack corresponding to the considerable amount of oxygen reduced by the cathodic reaction is concentrated on the microscopic area of steel exposed, and the *intensity of attack* (corrosion per unit area of the part affected) may be very great.

If, however, nickel, a metal only *slightly* more noble than iron, is used as a plating, corrosion, although it may occur at gaps, will not be intensified. The risk of gaps in nickel-plating declines as the coating-thickness is increased; it is false economy, therefore, to make the coating too thin (see ref. (8), p. 351). But even thick, non-porous nickel is not ideal since, in an atmosphere containing sulphur dioxide, it loses its brightness by formation of small amounts of a basic sulphate[22]. The phenomenon, known as 'fogging', is objectionable on aesthetic grounds, although not generally dangerous. It is today prevented by applying a much thinner coating of chromium outside the nickel. The finish commonly styled 'chromium-plating' generally consists of a layer of nickel covered with a flash of chromium; sometimes a layer of copper is included, either below the nickel-layer, or, in the so-called 'sandwich-system', dividing the nickel into two parts, so that the sequence is now

steel | nickel | copper | nickel | chromium

If the metal used for the covering is *less* noble than iron, the electric current will flow in the opposite direction. For instance, at a pore in a zinc coating on steel (Fig. 4.1(B)), the zinc will be the anode of the cell

steel | liquid | zinc

and will suffer attack, whilst the steel, being cathodic, receives protection. In the opening stages the protection will be 'sacrificial'; the zinc is attacked in averting the attack on the steel. Later, however, there is, in most liquids, cathodic deposition of zinc hydroxide on the steel and the current flowing decreases, for reasons analogous to those involved in inhibition by magnesium salts (p. 10). Thus after a time the attack on the zinc becomes slow, although it does not cease. The period over which a zinc layer provides protection (e.g. on a galvanized iron roof) is roughly proportional to the thickness of the zinc; here again, therefore, it is false economy to stint the amount of metal applied. For exposure to polluted atmospheres, it may be advantageous to paint galvanized surfaces after erection of the structure.

Despite its name, no galvanic current is employed in the manufacture of galvanized iron, which is produced by dipping steel into molten zinc. It is, however, possible to obtain a zinc coating by electro-plating; such coating can be bright and aesthetically more attractive than a galvanized coat, although, if thin, it is likely to protect only for a limited period. One of the best ways of using zinc to give protection likely to last many years is to apply a 'sprayed' coat by means of a pistol which distributes a shower of tiny globules of molten zinc on the steel surface; the globules flatten on striking the steel, and provide a porous coat with a structure resembling that of flaky pastry. The steel is generally roughened by shot-blasting before spraying, so that the applied zinc 'keys' into the holes in the surface and good adhesion is obtained.

The advantage of the spraying process is that it can be applied to a structure *after* erection. Thus weld-lines, rivets or bolts can be protected. Since sprayed coatings are porous, only metals anodic towards iron are, in general, suitable for this type of protection. Aluminium, as well as zinc, is frequently applied by spraying, and, under *severely corrosive* conditions, an aluminium coating will provide protection for a longer period than a zinc coating, since the aluminium, itself carrying a protective oxide-film, is attacked more slowly. Under certain *mild* conditions, the aluminium may be attacked too slowly to provide cathodic protection to the steel basis; in such circumstances zinc is preferred. Reference to Fig. 4.1(B) (p. 32) will make this clear. Here the zinc suffers rapid anodic attack, producing a movement of cations *towards* the exposed iron—i.e. in a direction opposite to that characteristic of corrosion; hence the iron is prevented from corroding. If zinc were to be replaced by something which was attacked with insufficient velocity, the cathodic current density (current per unit area of exposed iron) would be too small to afford protection.

However, the factors determining the choice of coating metal are complex, and it is well to consult an impartial but experienced person before making the decision.

4.3 Prevention of rusting by paint

One of the commonest ways of protecting the steel-work depends on paint. Here again it is possible to carry out the process *after* the erection of a structure, so that joints and crevices (the parts most liable to provide corrosion-cells) can receive thorough treatment; if two pieces of steel are to be joined by bolting or riveting, it is generally good practice to apply paint to the faying surfaces* *before* they are brought together, and to over-paint the whole external surface after erection.

Most commercial paint-coats are pervious to water vapour and oxygen. Nevertheless, by scientific choice of procedure, rusting can be prevented. The sensible plan is to cover the steel with a priming coat containing an inhibitive pigment, and then to add one or more outer coats chosen to provide mechanical protection (e.g. to prevent damage to the priming coat by particles of wind-borne grit); the outermost coat will, of course, be

---

* When two pieces of metal are held together by bolting or riveting, the surfaces which, if absolutely flat, would be in contact everywhere, are known as the *faying surfaces*. In practice, the surfaces are not absolutely flat and touch only at a few points. Elsewhere they are separated by a layer of air, or sometimes, under outdoor conditions, by water. This is obviously likely to set up corrosion, and the crevice should be filled up with paint, or with a stiff paste, known as a *jointing compound*.

chosen to provide the desired colour. An old-fashioned paint-system shown to be highly effective by comparative exposure tests is red lead in linseed oil as primer covered with two coats of iron oxide oil-paint; the pigment used in these outer coats may be the familiar red iron oxide, but for conferring robust properties the flaky form of ferric oxide known as micaceous iron ore has much to commend it. However, most of these old-fashioned paints based on linseed oil as vehicle dry too slowly to be popular today.

Chromate pigments in priming coats

The inhibition conferred by the priming coat probably depends on the general principles explained in connection with soluble inhibitors, but their working may be more complicated and in some cases there is still difference of opinion. Potassium chromate (probably for reasons suggested on p. 361 of ref. (8)) renders water non-rusting. To add potassium chromate to a paint would be useless; it would be soon washed away by the rain. However, certain zinc chromates are known which are sufficiently soluble to confer inhibition but not so soluble as to be quickly washed away. In practice, a mixture of iron oxide and zinc chromate is generally employed, dispersed in a mixture containing oil and resin; after the paint-coat has been applied to a steel surface, the organic mixture takes up oxygen from the air and polymerizes to give a solid film in which the two types of pigment are embedded.

Red lead in priming coats

One of the best inhibitive primers, used since the time of the Roman Emperor, Nero, is made from red lead. In past centuries, red lead, prepared by the controlled roasting of litharge (PbO) in air, always contained some residual PbO, probably at the centre of the larger grains, which consisted of Pb_3O_4 in their outer parts. When such red lead was rubbed with linseed oil and the resultant paint applied to steel without delay, excellent protection was obtained. If, however, painting was delayed, combination between PbO and fatty acids or other organic constituents of the oil would lead to the formation of soap-like substances, and the whole mixture would set to a solid mass in the paint-pot; the necessity for making up the paint freshly in small quantities, and of washing pots and brushes without delay was a great inconvenience. After much research, manufacturers succeeded in producing a pigment which was essentially Pb_3O_4 and which provided a non-setting paint; this was hailed as a triumph, but engineers were soon complaining that the new product did not give the excellent protection provided by old-fashioned red-lead paint. Comparative tests at Cambridge have confirmed this; paints made from non-setting red lead to which a little litharge had intentionally been added were, in fact, found to protect better than paints without added litharge[23].

It was found later[36] that the true inhibitors in red lead paints are

D

degradation products, such as the lead salt of azelaic acid, $COOH.(CH_2)_7$. COOH; this arises from the break-down of lead salts of linoleic and lino-lenic acids (both these acids are present as glycerides in linseed oil, and will rapidly form their lead salts if free PbO is present, but not if it is absent). Just how the azelate inhibits rusting is still uncertain, but it is believed that it accelerates the conversion of ferrous to ferric compounds close to the metal, so that a protective film is formed in union with the steel instead of non-protective rust.

Metallic lead in priming coats

Recently paints have come on the market containing metallic lead powder as pigment. Since lead oxide is absent, the objectionable 'setting in the pot' is avoided, but if the paint is spread out as a thin coating over a steel surface the oxygen of the air is likely to produce oxide, and setting can now take place; the formation of lead azelate, as in old-fashioned red-lead paints, also becomes possible. This may be the reason for the good protec-tion which seems to be obtainable from metallic lead paint, although other explanations have been offered. For many situations, however, the use of lead in paints is condemned on account of its toxicity.

Metallic zinc in priming coats

A different principle of protection is involved in the so-called *zinc-rich paints*. If metallic zinc is used in a paint with the volume-ratio of metallic pigment and organic vehicle adjusted so as to ensure that the metallic particles are in electrical contact with one another, the coating should provide protection to steel exposed at microscopic gaps by the same mechanism as was described in discussing coatings of metallic zinc (p. 33). The protection will be sacrificial at the outset, but later the deposition of zinc hydroxide on the steel should diminish the current flowing and thus the attack on the zinc should become slow. Only certain vehicles, such as polystyrene and chloro-rubber, will provide paints which possess the desired properties. Paints of this class have shown themselves capable of providing excellent protection—not as good, in general, as galvanized or sprayed zinc coatings, but often equal to red lead, and sometimes better. The main objection is that the circuit needed for cathodic protection may be broken if an outer coating consisting of non-conducting paint is applied on the zinc-rich layer. In cases where there is no imperative need for a coloured finish, it may be possible for the outer coating also to consist of zinc-rich paint, similar to the inner coating.

Need for cleaning of the surface before painting

Whatever paint-combination is chosen, it should be applied to a clean steel surface, free from 'mill-scale' (the relatively thick layer of oxide formed during the hot-rolling of steel) and also free from rust. Mill-scale is particularly objectionable if the steel has been exposed to the atmosphere for a few months before painting, since the scale is then likely to have flaked off on *small* areas; anodic attack occurs on the steel exposed on these areas, the scale-covered area being the cathode. The attack at the gaps in the scale may be intense, owing to the dangerous combination of large cathode and small anode (see p. 32). It is impossible to remove mill-scale by abrasion with emery-paper, although such treatment may 'burnish' the scale and give a false impression that the surface exposed consists of clean metal. Mill-scale can, however, be removed by grit-blasting or by 'pickling' in acid.

Rust, if free from salts, is relatively harmless, but in general rust produced in urban or industrial conditions will contain ferrous sulphate, whilst that produced near the sea will contain ferrous chloride. If paint is applied over rust containing nests or patches of ferrous salts, further local rust-production will take place below the paint at these points, since ferrous salts in rust will stimulate rusting, as shown on p. 21. If this fresh rust is formed locally (at the patches of ferrous salt), it will occupy a larger volume than the iron destroyed in producing it; consequently, the paint-coat will locally be humped up, and, being incapable of stretching to a sufficient extent, it will crack, leaving the steel surface unprotected. Thus paint applied to a surface carrying typical rust provides protection for a period much shorter than paint applied to a clean surface. The removal of rust is best carried out, like the removal of mill-scale, by grit-blasting or pickling; but, if mill-scale is absent, vigorous abrasion with emery cloth is much better than nothing for the removal of the rust.

5 Closing remarks

As stated earlier, there is no need to abandon the use of steel on account of its liability to rust. Sometimes it may save money and trouble in the long run to replace steel by a more costly material which requires less maintenance. In other instances, the expense of replacement could be avoided by suitable protective measures.

Salesmen advocating the adoption of certain non-ferrous materials base arguments on the fact that such materials do not rust; that is true, but it is not the same thing as saying that these materials do not corrode. The attack, although generally less disfiguring than that on steel, is not necessarily of a kind that can be neglected.

Aesthetic conditions apart, it could perhaps be argued that the conspicuous character of rust may sometimes be an advantage; the red colour is a danger-signal calling attention to the fact that attack is taking place and this signal is given at an early stage when, on a non-ferrous metal, the corresponding quantity of a white corrosion-product would pass unnoticed. Also, the rusting of steel generally occurs on the exposed part of the surface and is thus conspicuous, whereas on some resistant materials the visible outer surface remains unchanged, but corrosion-products are formed unseen in crevices where the protective oxide-film is not kept in repair; the corrosion-products may be voluminous and capable of causing disruption by leverage. There is no desire to underestimate the value of non-ferrous materials, which are playing an essential part in the fight against Corrosion. But the arguments are not all on one side.

If the protective measures for steel are adopted, extreme care is needed in their application. Particularly, as already stated, the surface must be suitably prepared before a protective coating is spread or deposited on it. This was known at the start of the century, and it seems possible that practical engineers were then better informed about the dangers of mill-scale and rust below paint than some of their successors of the present day. The addition of new teaching subjects to the engineering curriculum at colleges inevitably tends to force out the old ones, so that today in some engineering departments little attention seems to be given to Corrosion. For that reason anyone who intends to enter the Engineering Profession may with advantage start his study of Corrosion at the school stage; having acquired interest in the subject, he can keep up with the advances of knowledge by private reading—perhaps becoming a member of the British Joint Corrosion Group, on which the four important Engineering Institutions are represented. If, at the college stage, he receives formal instruction in the subject —so much the better; but if he receives none, he will not be entirely helpless when in later life he finds himself faced with practical problems connected with Corrosion; he can for instance consult some of the books mentioned in the Bibliography printed on page 49.

Appendix: Experimental Work

(Adapted from *Practical Experimental Metallurgy* by D. Eurof Davies (1966) Elsevier Publishing Company. Chapters 14 and 15)

These experiments demonstrate the electrochemical nature of corrosion of iron in salt solution, i.e. corrosion involves an anode, a cathode and a conducting electrolyte. At the cathode, oxygen is taken up giving sodium hydroxide and at the anode, metal dissolves.

EXPERIMENT 1   ANODIC AND CATHODIC AREAS SET UP BY DIFFERENTIAL AERATION

The distribution of corrosion produced by drops of sodium chloride solution placed on a horizontal sheet of iron is due to differential aeration currents. These currents are set up by the difference in oxygen concentration at two places on a metal surface, the place where there is excess oxygen being cathodic. A drop of sodium chloride solution resting on iron sets up attack at the centre where ferrous chloride is formed, whilst the peripheral zone remains immune from attack, sodium hydroxide being there produced; a ring of yellow rust appears where the alkali and iron salt meet and interact (Fig. A1.1).

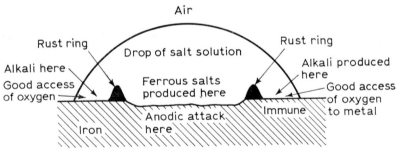

FIG. A1.1   The action of a drop of sodium chloride on iron.

This distribution is connected with electric currents flowing between a central anode and a cathode at the periphery, where oxygen, the cathodic reactant, can best be replenished (Fig. A1.2).

When the sodium chloride drops contain small amounts of ferroxyl indi-

40

cator (phenolphthalein and potassium ferricyanide) the cathodic places are shown up as pink patches and the anodic places (where iron salts appear) as blue points. If the solution contains oxygen at the outset, the primary distribution (Fig. A1.3a) is an irregular pattern of blue and red.

When the original oxygen has been used up, the red colour becomes confined to the peripheral zone, since only here can oxygen be readily replenished. Thus the secondary distribution (Fig. A1.3b) will be a pink ring separated from a blue centre by a brown circle of rust. The time needed to pass from the primary to secondary distribution naturally depends on the original oxygen concentration; if the liquid is supersaturated with oxygen the primary distribution remains abnormally long, whereas if dissolved oxygen is removed before the drop is placed on the metal, the secondary distribution appears almost immediately.

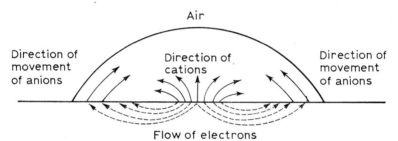

FIG. A1.2 Electrical phenomena in corrosion.

Experimental instructions

Prepare an agar-agar gel containing 3% NaCl by dissolving 7.5 g of sodium chloride in 250 cm³ distilled water. Add 5 g of powdered agar-agar and boil the solution until the agar is dispersed. Now add 5 cm³ of a 5% potassium ferricyanide (hexacyanoferrate(III)) solution and 1 cm³ of a 1% solution of phenolphthalein. This mixture of potassium ferricyanide and phenolphthalein is called the ferroxyl indicator. Use 45 cm³ of this solution for each experiment and record your observations illustrating them by sketches in each case. Abrade a piece of mild steel and then clean it with acetone. Place a large drop of the gel on the metal surface and observe the sequence of changes.

Although no thick oxide film is present on the specimen surface after abrasive treatment, a thin invisible film is formed which keeps breaking at places where the stresses left in the metal by the abrasion are tensional. Thus a series of anodic points appear along lines parallel to the abrasion direction, shown up by blue spots; at the same time, pink coloured cathodic areas will develop. In time, however, the blue points around the edge of the drop disappear, and the pink regions unite to form a complete ring. In the centre of the drop the blue regions unite to give a central spot. Separating the blue and pink regions is a ring of brown rust.

The cathodic reaction can only proceed at places where oxygen is available. At the centre of the drop the oxygen dissolved in the solution is soon exhausted, thus the cathodic reaction stops, but at the edge of the drop fresh oxygen may be taken into solution from the atmosphere and the cathodic reaction can continue. Any ferrous chloride produced in the peripheral region is in an alkaline environment; consequently hydrated ferric hydroxide will be precipitated in physical contact with the metal, so stifling the anodic reaction. In the centre of the drop, the anodic reaction may continue, since the solution is not alkaline, and no precipitate forms. At the interface between the alkali ring and the ferrous chloride central spot, ferrous hydroxide is precipitated and is quickly oxidized to hydrated oxide (brown rust) which can be written $Fe_2O_3.H_2O$ or $FeO.OH$.

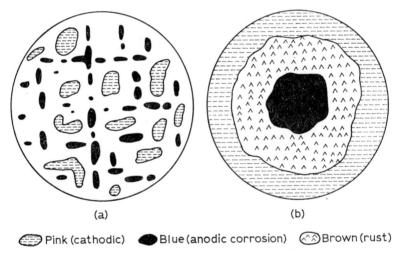

(a)   (b)

⟨⟩ Pink (cathodic)   ● Blue (anodic corrosion)   ⟨∧∧⟩ Brown (rust)

FIG. A1.3 Use of ferroxyl indicator to investigate sodium chloride drop corrosion. (a) Primary (b) Secondary distribution of cathodic and anodic areas.

EXPERIMENT 2   ANODIC AND CATHODIC AREAS SET UP BY GALVANIC COUPLING

In order for corrosion to occur, there must be a potential difference. This can be set up due to differential aeration as already discussed. A potential difference also exists when two dissimilar metals are connected together in an electrolyte. This difference in potential enables a current to flow between one metal as a cathode and the other as an anode. The resulting attack is known as galvanic corrosion. When galvanic corrosion is considered it is essential to know which metal will be the anode, because this is the one which will corrode. A rough indication is given by the positions of the metals in the standard electrochemical series (Table A1), although, strictly speaking, the potential values refer to film-free metal standing in a normal solution of ions of the same metal.

In the standard electrochemical series, as the metals become more positive than hydrogen, they become very inert, e.g. platinum is not attacked by nitric acid. Similarly, in the negative direction, the metals become increasingly reactive, for instance, magnesium displaces hydrogen from a neutral salt solution such as sodium chloride.

When two dissimilar metals are combined together in an electrolyte, the one near the top of the series will generally be the anode and will suffer accelerated corrosion in a galvanic couple, whilst the more positive metal will be the cathode which does not corrode. Thus, in a copper/iron couple, copper is the cathode and iron the anode. With iron and zinc on the other hand the iron is the cathodic member and does not corrode. This is an example of cathodic protection, and is used for protecting iron against corrosion as in a buried line, where magnesium anodes are connected to the pipe line at various intervals. Ships' bottoms are also protected in this way, zinc or magnesium anodes being welded on to the steel plates. Other examples are galvanized buckets and sheets.

TABLE A1. STANDARD ELECTROCHEMICAL SERIES

Metal	*Electrode potential(Volts)*
Magnesium	−1.80*
Zinc	−0.77
Iron	−0.44
Cadmium	−0.42
Nickel	−0.25
Tin	−0.14
Lead	−0.13
Hydrogen	0.00
Copper	+0.35
Silver	+0.80
Platinum	+1.20

Cold-working also causes a change in the electrode potential. The shift is not always in the same direction, but often the cold-worked part is anodic to the rest. The ends of a nail which have been deformed by cold working will be anodic to the shank which has not been so severely deformed.

Experimental instructions

The existence of the anodes and cathodes, whether they be due to differential aeration, dissimilar metals, or a difference in cold work, may be demonstrated by means of the ferroxyl indicator.

(1) In order to demonstrate the accumulation of alkali at the cathodic

* This value cannot be obtained experimentally, owing to local reactions such as the 'evolution of hydrogen' which complicate the situation.

areas and corrosion at the anodic areas, fill a Petri dish with 45 cm³ of the hot gel and allow to cool without disturbance until it begins to set. At this point place a partially copper-plated nail, i.e. one with one half of its length plated with copper, in the gel and observe the subsequent colour developments over a period of several hours. The red colour arises from the accumulation of alkali on the cathodic areas and the blue colour reveals the presence of ferrous ions at the iron anode surface.

(2) Repeat the experiment with a nail partially plated with zinc. The faint white colour around the zinc is due to zinc ferricyanide.

(3) In order to demonstrate the existence of, and to show the location of anodes and cathodes on a single metal surface, repeat the experiment with a plain nail. The red colour is not so noticeable this time.

(4) Connect a bare iron nail to a strip of zinc by means of a thin wire making sure of a good electrical contact. Bend the zinc strip so that it will rest in a Petri dish without support. Then pour the warm gel solution into a Petri dish and allow to cool. As the gel begins to set, place the nail and zinc strip in the gel and examine periodically. This is an example of corrosion prevention by means of cathodic protection.

(5) Repeat the previous experiment using a copper-plated nail in place of the iron nail. The iron receives cathodic protection from the zinc despite the presence of the copper/iron couple.

EXPERIMENT 3   CORROSION INHIBITORS

Corrosion is an electrochemical phenomenon involving anodes and cathodes, and is dangerous because it frequently leads to the formation of soluble bodies at these anodes and cathodes. Under these circumstances the reactions which take place cannot be stifled. If on the other hand a sparingly soluble body forms in contact with the metal, then the corrosion stops, e.g. lead in sulphuric acid. The corrosion product is insoluble lead sulphate in contact with the lead which prevents further attack by isolating the metal from sulphuric acid. However, in most cases the corrosion products are soluble as in the case of iron in a sodium chloride solution.

At the anode ferrous ions form
$$Fe = Fe^{2+} + 2e$$
At the cathode oxygen will be reduced
$$\tfrac{1}{2}O_2 + H_2O + 2e = 2(OH)^-$$
Since sodium ions and chloride ions are already present in the solution, the cathodic product can be regarded as sodium hydroxide and the anodic product as ferrous chloride. Both of these are freely soluble bodies and will not stifle attack. They will, however, yield solid ferrous hydroxide where they meet
$$FeCl_2 + 2NaOH = Fe(OH)_2 + 2NaCl$$
but if oxygen is present in sufficient quantity, the ferrous hydroxide will be

immediately oxidized to ferric hydroxide—the substance commonly known as brown rust—$Fe_2O_3.H_2O$ or $FeO.OH$.

Thus in electrochemical corrosion, the iron goes into solution at one place, oxygen is taken up at a second place and hydrated oxide is formed at a third point, and since it is produced away from the anodic region, which is the point of attack, it cannot stifle the corrosion.

Chemicals which, when added to a corrosive solution, check the corrosion of the metal or prevent it altogether, are known as inhibitors. These inhibitors produce insoluble corrosion products. When the anodic reaction is stifled, it is called anodic inhibition. Cathodic inhibitors stifle the cathodic reaction. Examples of anodic inhibitors are: sodium phosphate, sodium carbonate and sodium bicarbonate. Zinc sulphate and the salts of calcium and magnesium are cathodic inhibitors. Chromates are powerful inhibitors but are considered by different authorities to interfere with the anodic or cathodic reactions—or both. A quantity of an inhibitor which, added to pure water would render it non-corrosive, may fail to do so, if sodium chloride is already present. In general, the amount of inhibitor

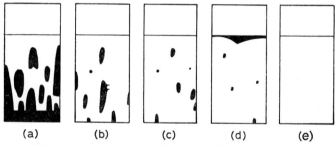

(a)   (b)   (c)   (d)   (e)

FIG. A2.1 Effect of increasing quantities of inhibitor converting (a) General attack through (b) Restricted attack to (c) Localized attack, then (d) Water-line attack and finally (e) Immunity.

needed increases with the chloride concentration. The inhibitor must be present in sufficient quantity so that it can diffuse to the anodic areas more quickly than ferrous ions can be produced by the electro-chemical corrosion process. If the anodic product is converted to phosphate or carbonate at a distance from the point of attack, there can be no stifling. Similarly for a cathodic inhibitor, whilst the direct deposition of zinc hydroxide on the cathodic surface will diminish the rate of attack, this will not happen if sodium chloride is present at such a concentration that zinc hydroxide is only precipitated at a distance from the surface.

Anodic inhibitors are efficient, but dangerous, because if added in quantity just insufficient to abolish corrosion completely they localize the attack and render it more intense than if no inhibitor had been introduced, e.g. steel immersed in a chloride/carbonate solution. When sodium carbonate is added in increasing amounts to a sodium chloride solution, the area of corrosion contracts, but, just as corrosion seems to disappear altogether, a new zone of intense attack appears along the water line. Still higher concentration of inhibitor may prevent the attack entirely (Fig. A2.1).

Similarly, when a drop of sodium chloride solution containing an anodic inhibitor, just insufficient to prevent attack completely, is placed on the steel, corrosion will start along the drop boundary—the place which, with sodium chloride solution alone, would be immune. As the concentration of anodic inhibitor increases the total corrosion decreases, but the intensity of corrosion increases, expecially at the drop boundary, until the concentration is such as to prevent corrosion altogether.

Experimental instructions

Degrease a piece of steel (7.5 × 2.5 × 0.4 cm) by cleaning with cotton wool soaked in acetone, and then abrade the surface thoroughly with emery paper. Repeat the cleaning, and thereafter handle the specimens only with tongs to avoid contaminating the surface. Repeat this procedure also with fourteen other steel specimens (7.5 × 2.5 × 0.4 cm) each of which has a hole of about 3 mm drilled at one end for suspensions.

Prepare 150 cm³ of each of seven chloride/chromate mixtures and also 150 cm³ of each of seven chloride/carbonate mixtures, of various concentrations as follows. Table A2 shows the ingredients of the first seven solutions. To make solution 1, for example (see Table A2), mix, in a 250 cm³ beaker, 150 cm³ of 0.1 M NaC1 and 0.45 cm³ of M K_2CrO_4. The resulting solution is approximately 0.1 M with respect to sodium chloride and 0.003 M with respect to potassium chromate. Solutions 2 to 7 are prepared in similar fashion. The solutions (8 to 14) of the chloride/carbonate mixtures are likewise prepared, using the quantities listed in Table A3.

Lay the large steel sheet on a glass plate and place three drops of each of the fourteen solutions on it and also some drops of pure 0.1 M NaCl solution. Cover the steel sheet with an inverted glass vessel, sealing the edges

Table A2 Preparation of chloride/chromate mixtures

Solution number	Volume of 0.1 M NaCl /cm³	Volume of MK_2CrO_4 (in cm³) or mass of solid K_2CrO_4 (in g)	Approximate molarity of solution with respect to sodium chloride	Approximate molarity of solution with respect to potassium chromate
1	150	0.45 cm³	0.1M	0.003M
2	150	0.75 cm³	0.1M	0.005M
3	150	1.5 cm³	0.1M	0.01M
4	150	2.25 cm³	0.1M	0.015M
5	150	4.5 cm³	0.1M	0.03M
6	150	2.913 g	0.1M	0.1M
7	150	8.739 g	0.1M	0.3M

of the vessel with damp cotton wool to prevent evaporation of the drops.

Also, pour out 130 cm³ of each of solutions 1 to 14 into fourteen 150 cm³ beakers labelled 1 to 14. Weigh each of the fourteen steel specimens and partially immerse them in the fourteen solutions, so that specimen 1 is immersed in solution 1 in beaker 1, specimen 2 is immersed in solution 2 in beaker 2, and so on.

Table A3 Preparation of chloride/carbonate mixtures

Solution number	Volume of 0.1M NaCl /cm³	Volume of 3M Na_2CO_3 (in cm³) or mass of solid Na_2CO_3 (in g)	Approximate molarity of solution with respect to sodium chloride	Approximate molarity of solution with respect to sodium carbonate
8	150	0.25 cm³	0.1M	0.005M
9	150	0.5 cm³	0.1M	0.01M
10	150	1.5 cm³	0.1M	0.03M
11	150	3.0 cm³	0.1M	0.06M
12	150	5.0 cm³	0.1M	0.1M
13	150	4.77 g	0.1M	0.3M
14	150	7.95 g	0.1M	0.5M

Suspend the specimens in the solution by passing a glass rod through the 3 mm hole and letting the ends of the glass rod rest on the rim of the beaker. Do not allow the steel specimens to touch the beaker at any time.

Record the changes which have taken place in both drops and specimens after forty eight hours. Then clean the specimens under a tap, dry, and re-weigh. Draw sketches showing the distribution of the corrosion and explain your results.

TA
467
.E92

17,291

Bibliography

I Books and Journals

(1) BÉNARD, J., *L'oxydation des Métaux*, Gauthier-Villars, 2 vols, 1962, 1964.

(2) BIANCHI, G., and MAZZA, F., *Fondamenti di Corrosione e Protezione dei Metalli*, Tamburini, Milan, 1968.

(3) BUTLER, G., and ISON, H. C. K., *Corrosion and its Prevention in Waters*, Leonard Hill, 1966.

(4) CHILTON, J. P., *Principles of Metallic Corrosion*, Royal Institute of Chemistry, 1961.

(5) EUROF DAVIES, D., *Practical Experimental Metallurgy*, Elsevier, 1966, Chapters 13 to 19.

(6) EVANS, U. R., *Corrosion of Metals*, Arnold, 2nd edition, 1926.

(7) EVANS, U. R., *Corrosion and Oxidation of Metals*, Arnold, 1960.

(8) EVANS, U. R., *Corrosion and Oxidation of Metals:* first supplementary volume, Arnold, 1968.

(9) EVANS, U. R., *Introduction to Metallic Corrosion*, Arnold, 2nd edition, 1963.

(10) HUDSON, J. C., *Corrosion of Iron and Steel*, Chapman and Hall, 1940.

(11) KAESCHE, H., *Die Korrosion der Metalle*, Springer, 1966.

(12) KUBASCHEWSKI, O., and HOPKINS, B. E., *Oxidation of Metals and Alloys*, Butterworths, 2nd edition, 1962.

(13) LAQUE, F. L., and COPSON, H. R., *Corrosion Resistance of Metals and Alloys*, Reinhold, 2nd edition, 1964.

(14) National Chemical Laboratory: Annual Reports 1959, 1962, 1964.

(15) POURBAIX, M., *Atlas of Electrochemical Equilibria in Aqueous Solutions*, Pergamon, 1966.

(16) ROGERS, T. H., *Marine Corrosion*, Newnes, 1968.

(17) SCHIKORR, *Häufige Korrosionsschaden an Metallen und ihreVermeidung*, Wittwer, Stuttgart.

(18) SHREIR (editor), *Corrosion*, Newnes, 2 volumes, 1963.

(19) SPELLER, F. N., *Corrosion: causes and prevention*, McGraw-Hill, 3rd edition, 1951.

(20) TOMASHOW, N. D., *Corrosion of Metals and Alloys*, Moscow: English translation by A. D. Mercer; editor G. J. L. Booker (National Lending Library, Boston Spa).

(21) UHLIG, H. H., *Corrosion and Corrosion Control*, Wiley, 1963.

(22) VERNON, W. H. J., Jubilee and Cantor Lectures, reprinted in *Chem. and Ind.*, 1943, p. 314; *J. roy. Soc. Arts*, 1948–49, **97**, 578, 593; for data on 'Conservation of National Resources' see *Instn. Civ. Engrs.*, 1956–57, p. 105, esp. p. 130.

II Papers in Journals

(23) BRITTON, S. C., and EVANS, U. R., *J. Soc. chem. Ind.*, 1939, **58**, 90.
(24) BUKOWIECKI, A., *Schweizer Archiv angew. Wiss.*, 1957, **23**, 97.
(25) CHANDLER, K. A., and KILCULLEN, M. B., *Brit. Corr. J.*, 1968, **3**, 80.
(26) COPSON, H. A., *Proc. Amer. Soc. Test. Mat.*, 1945, **75**, 554; 1960, **60**, 650.
(27) EDELEANU, C., and EVANS, U. R., *Trans. Faraday Soc.*, 1951, **47**, 1121.
(28) EVANS, U. R., *J. Soc. chem. Ind.*, 1924, **43**, 315T.
(29) FORREST, H. O., ROETHELI, B. E., and BROWN, R. H., *Ind. engng. Chem.*, 1931, **23**, 650.
(30) GILROY, D., and MAYNE, J. E. O., *Corrosion Science*, 1965, **5**, 55.
(31) HUDSON, J. C., and STANNERS, J. F., *J. appl. Chem.*, 1953, **3**, 86.
(32) KENWORTHY, L., *J. Inst. Met.*, 1943, **69**, 67.
(33) LORKING, K. F., and MAYNE, J. E. O., *J. appl. Chem.*, 1961, **11**, 170.
(34) MAYNE, J. E. O., *J. chem. Soc.*, 1953, p. 129; *J. Iron Steel Inst.*, 1954, **176**, 140; *J. appl. Chem.*, 1959, **9**, 673.
(35) MAYNE, J. E. O., and MENTER, J. W., *J. chem. Soc.*, 1954, p. 99.
(36) MAYNE, J. E. O., and VAN ROOYEN, D., *J. appl. Chem.*, 1954, **4**, 384.
(37) PRESTON, R. ST. J., and SANYAL, B., *J. appl. Chem.*, 1956, **6**, 26.
(38) SCHIKORR, G., *Werkstoffe u. Korrosion*, 1963, **14**, 69; 1967, **18**, 514.
(39) TURNER, M. E. D., *Chem. and Ind.*, 1963, p. 517.
(40) VERWEY, E. J. W., *Bull. Soc. Chim. France*, 1949, p. D120.

(Some of the journals quoted above will be found in any large scientific library. If a paper cannot be consulted directly, a summary may in some cases be found by looking up the Author's name in the index of a textbook, e.g. one of those quoted in Refs. (1), (7), (8), (12) and (18).)